Management of
Industrial Maintenance

A. KELLY
MSc(Eng), C Eng, MIMech E

and

M. J. HARRIS
MSc, MInst P

Senior Lecturers in Engineering,
Simon Engineering Laboratories,
Victoria University of Manchester

NEWNES - BUTTERWORTHS
LONDON - BOSTON
Sydney - Wellington - Durban - Toronto

The Butterworth Group

United Kingdom	**Butterworth & Co (Publishers) Ltd** London: 88 Kingsway, WC2B 6AB
Australia	**Butterworths Pty Ltd** Sydney: 586 Pacific Highway, Chatswood, NSW 2067 Also at Melbourne, Brisbane, Adelaide and Perth
Canada	**Butterworth & Co. (Canada) Ltd** Toronto: 2265 Midland Avenue, Scarborough, Ontario M1P 4S1
New Zealand	**Butterworths of New Zealand Ltd** Wellington: T & W Young Building, 77–85 Customhouse Quay, 1 CPO 472
South Africa	**Butterworth & Co (South Africa) (Pty) Ltd** Durban: 152–154 Gale Street
USA	**Butterworth (Publishers) Inc** Boston: 10 Tower Office Park, Woburn, Mass. 01801

First published 1978
reprinted 1979
© Butterworth & Co (Publishers) Ltd, 1978

British Library Cataloguing in Publication Data

Kelly, A
   The management of industrial
   maintenance. – (Management library).
   1. Plant maintenance – Management
   I. Title II. Harris, M J III. Series
658.2'02   HD69.M3   77–30308

ISBN 0-408-00297-2

Typeset by Butterworths Litho Preparation Department

Printed in England by Billing & Son Limited,
Guildford, London and Worcester

Preface

Industrial plant is now more automated, much larger, and very much more expensive to own than ever before. Because of the high cost of maintenance resources, and of equipment availability, maintenance effort has become a critical factor in company profitability. In addition, the growth in sophistication and in the technological content of maintenance work has made its management far more difficult. An inevitable corollary of all this is that industry should devote as much forethought to the planning of the maintenance function as, say, to the production function. Unfortunately this has not happened and, with the exception of a few large process companies, maintenance remains the Cinderella function of British industry. The situation has not been helped by the failure of universities and polytechnics to involve themselves at all significantly in teaching and research in this area, a state of affairs reflected in the dearth of books treating maintenance engineering and management at a high (but not necessarily mathematical) level *and* in a practically orientated way. The contrast with the literary output in production engineering and management makes this clear.

This book has been written to fill the obvious gap in the technical literature. It has been designed to fit the needs of the practising professional engineer, directly or indirectly involved in industrial maintenance, and also of the student of industrial management at university or polytechnic. It has been written as a result of:

1. Post-experience courses directed by the authors on behalf of the Institution of Mechanical Engineers.

2. Research and consultancy projects under the auspices of the Engineering Department of Manchester University.

3. The MSc course in terotechnology which was started at Manchester in 1976.

Consequently, several chapters are contributions by practising engineers and much of the other material has arisen from analysis of the industrial maintenance situation.

Chapter 1 reviews the overall operation of the maintenance function in terms of the causes, characteristics and costs of maintenance work. This introduction emphasises the importance of reliability and maintainability to the maintenance work load, topics which are covered in some detail in Chapter 3. Chapter 2 looks at decision making, failure statistics and the use of the latter in failure cause diagnosis. Planning and organisation of maintenance work are discussed in Chapters 4 and 5, problem solving by computerised simulation in Chapter 6. Chapters 7 and 8 complement Chapter 5 in describing, respectively, the organisation of spares and of network planning. One of the most important sections is Chapter 9, an extensive review of condition monitoring, which looks like becoming a cornerstone of maintenance planning. Chapters 10 and 11 review those management and engineering techniques, such as work measurement and logical fault finding which, over the last decade, have come into standard use in maintenance departments. Finally, the case studies of Chapters 12 and 13 illustrate the practical application of much of the material that has gone before.

To summarise, the aim of this book is to assist practising and potential engineering managers towards a fundamental understanding of the industrial maintenance function and the techniques available for organising and controlling it. A professional approach to maintenance is vital if management is to control plant and not the reverse.

A.K.
M.J.H.

Acknowledgements

We are deeply indebted to colleagues in the industrial and academic world who most generously contributed several important sections of our book. These sections, and their authors, are as follows:

Section 3.5 Case Study 1: A plant availability assessment.
Bob Moss, National Centre of Systems Reliability.
Chapter 8 Network analysis.
Harry Moody, ICI (Organics) Ltd.
Chapter 9 Condition-based maintenance.
Tim Henry, Simon Engineering Laboratories.
Chapter 12 Case Study 2: Reorganisation of a maintenance trade force.
Roy Warburton, BICC Ltd.

Those who have also contributed, either through collaboration in our various courses and research projects or through discussions and correspondence, are:

Derek Arstall, Lankro Chemicals Ltd.
Ben Blanchard, Prof., Virginia Polytechnic Institute
Bill Geraerds, Prof., Eindhoven Technical University
Bill Gore, Shell Petrochemicals (UK) Ltd.
Sid Howarth, ICI (Paints) Ltd.
David Mayers, CEGB (SE Region)
Harry Riddell, ICI (Organics) Ltd.
Knut Swärd, ALI RATI (Sweden)
Bernard Wilson, Lever Bros Ltd.
Peter Wood, ICI (Organics) Ltd.

and the following former students of mechanical engineering at the Simon Engineering Laboratories:

David Brockington, Gordon Cook, John Halstead and Phil Tempest.

We would like to single out, for our especial thanks, Dennis Parkes, Director of the National Terotechnology Centre, and Keith Lewis, Lecturer at the Simon Engineering Laboratories, the former for his general encouragement and the latter for his stimulating and positive criticism of our work.

A.K.
M.J.H.

Contents

Chapter 1

Maintenance Management, a Review

1.1 Maintenance and Profitability

Industrial organisations exist to make a profit; they use equipment and labour to convert raw materials into finished goods of higher value.

In the simplest terms the profit is the difference between the *income* from the sale of the product and the *costs* of the manufacture and sale of the product. Costs can be classified as fixed (e.g. cost of equipment and buildings) or variable (e.g. cost of raw materials). Profitability is influenced by many factors, e.g. customer demand, product price, equipment output, equipment capital cost and life, equipment running cost, etc. Maintenance is related to profitability through equipment output and equipment running cost. Maintenance work raises the level of equipment performance and availability but at the same time it adds to running costs. The objective of an industrial maintenance department should be the achievement of the optimum balance between these effects, i.e. *that balance which maximises the department's contribution to profitability*.

Over the last decade the dependence of profitability on maintenance effort has greatly increased. This is because industrial plant has become larger, downtime costs greater, and maintenance work more sophisticated and costly. In 1971, attention was drawn to this point in a Ministry of Technology Working Party report[1]; it was estimated that maintenance was costing the U.K. some £M3000 per year. The working party considered that substantial savings could be made by improving maintenance management and also by paying greater attention to factors affecting maintenance at other stages in the equipment life cycle. This 'life cycle' approach to maintenance cost reduction has since been defined as *terotechnology*[2].

1

1.2 Maintenance Management and Terotechnology

The scope and interrelationships of terotechnology are shown in *Figure 1.1*. During its life cycle, industrial equipment can be considered as passing through a number of stages, the first being design and the last replacement[3]. The level of maintenance required at the equipment operation stage is affected by factors at other stages.

At the design stage reliability and maintainability are the important factors and must be considered in relation to equipment performance, capital cost and running cost. At the installation stage maintainability continues to be an important factor because it is only then that the multidimensional nature of many of the maintenance problems becomes clear. The commissioning stage is not only a period of technical performance testing but also a learning period where primary design faults that might affect equipment availability are located and designed out. Finally, throughout the whole operational life a suitable learning system should be continued.

The function of a learning system is to gather and provide information on maintenance problem areas, thus facilitating determination of the plant's optimum maintenance operation. Since the design of equipment is a continuing process, information thus gathered should, ideally, be continuously fed back to the equipment manufacturer and in certain circumstances to a data bank which could be shared on an inter-company, national or international basis[4]. The difficulties of these last operations continue to pose a major obstacle to the successful implementation of a terotechnological approach; communication systems are expensive and different organisations (with different objectives) are involved during the equipment life cycle.

This book is not concerned with terotechnology as a whole but with maintenance management, a vital component which deals mainly with the equipment operation stage. Maintenance management can be considered as the direction and organisation of resources in order to control the availability and performance of industrial plant to some specified level. The maintenance manager has two main problems — the determination of the size and nature of the maintenance work load and the organisation and control of men, spares and equipment to meet this work load. In this book these problems will be considered in depth. Two preliminary exhortations are, however, required:

1. The equipment user must cooperate with the designer-manufacturer-installer in a full analysis of reliability and maintainability factors[5]. The higher the costs (production-lost costs and resource costs) of maintenance the more vital is cooperation; improvements

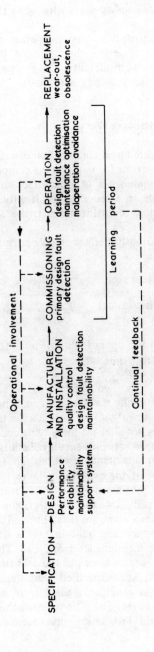

Figure 1.1. The equipment life cycle and factors affecting maintenance costs

in reliability and in maintainability are costly when the equipment is in use.

2. The maintenance department must cooperate closely with the production department in achieving the optimum balance between maintenance resource costs and availability. This is especially important where unavailability costs change greatly with time.

1.3 The Function of Maintenance Work

Maintenance can be considered as a combination of actions carried out in order to replace, repair, service (or modify) the components, or some identifiable grouping of components, of a manufacturing plant so that it will continue to operate to a specified availability for a specified time. In short, the function of maintenance is the control of plant availability.

Availability has been variously defined, with perhaps the most basic definition[6] being

$$\text{Availability (over a specified time)} = \frac{T_{up}}{T_{up} + T_{down}}$$

where

T_{up} = Cumulative time of operation in the nominal working state
T_{down} = Cumulative outage time.

This definition presumes only two defined states, namely working (1) and failed (2), whereas usually there will of course be a spectrum of intermediate states. Amendment of the formula to take account of this is not difficult and does not alter the underlying concept.

The maintenance manager can influence availability only through those outages which occasion maintenance actions. Thus T_{down} must be modified[3] to allow for this if the availability index is to be used for maintenance management purposes. *Figure 1.2* outlines the factors which affect such outages and shows their connection with availability. Of these factors the most important are reliability and maintainability. These are characteristics which are built in at the design stage and which thereafter affect the maintenance work load. The reliability of a unit (see Chapter 3) has been defined[7] as the probability that it will perform a specified function, under specified conditions, for a specified time. In some situations, for example, a measure of reliability is the mean running time to failure (MTTF). The maintainability of a unit (a characteristic of design and installation) has been defined[3] as the

5

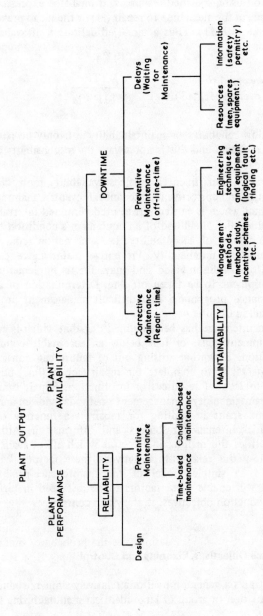

Figure 1.2. Factors influencing availability

probability that it will be restored to a specified condition within a given time period using specified resources. It might be expressed, for example, in terms of the mean time to repair ($\overline{M}$) or the mean preventive maintenance time (MPMT). Thus a modified definition of availability can be written as

$$\text{Availability (over a specified time)} = \frac{\text{MTTF}}{\text{MTTF} + \overline{M}}$$

The causes of low reliability or maintainability can only be removed by engineering re-design and this is not always the responsibility of the maintenance department.

The maintenance department influences availability more directly through preventive and corrective maintenance. Preventive maintenance is defined[8] as that which is carried out at predetermined intervals and intended to reduce the likelihood of an equipment's condition falling below a required level of acceptability; i.e. such action reduces the effects of equipment unreliability. Preventive maintenance can be time-based and/or condition-based and may, for its implementation, require the equipment to be taken off line. Determination of a preventive maintenance programme is a difficult management problem, discussed in detail in Chapter 4.

Corrective maintenance has been defined[8] as that which is carried out when equipment fails, or falls below an acceptable condition, while in operation. Downtime arising out of failure can consist not only of the time taken to complete the repair (repair time) but also of delays due to lack of resources or information. Repair time is a function of maintainability, management methods and engineering techniques. Time spent in waiting for repair is a function of the organisation of maintenance resources and information. Although *Figure 1.2* identifies the maintenance factors which affect availability it does not show the relationships between these factors. What is now required is a simple model of the maintenance-production situation which will enable these factors to be discussed in terms of maintenance-production objectives, strategy and control.

1.4 Maintenance Objectives, Planning and Control

Figure 1.3 illustrates a general model of a relatively simple production-maintenance situation (a group of large identical manufacturing units)

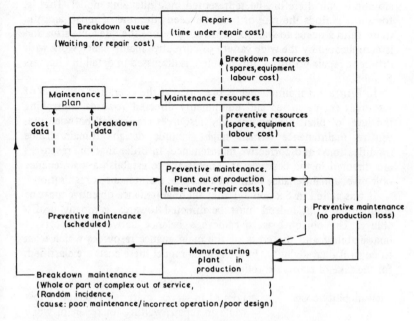

Figure 1.3. A general model of a maintenance system

which contains the essential characteristics of maintenance work. The units are considered to be in one of a number of states:

In production	**AVAILABLE**
Undergoing repair, waiting for repair, undergoing preventive maintenance	**UNAVAILABLE**

Movement from the available state to the unavailable state is caused by the need for preventive or corrective maintenance.

Preventive maintenance is carried out to control the level of failures but sometimes its execution requires the plant to be taken off line, i.e. to be made unavailable.

Since the degree of reliability is a cost compromise made at the design stage, a certain level of failure, and hence unavailability, must be expected. In practice, however, the level of failure is often much higher than expected (or acceptable) and it has been shown[9] that the additional failures are due either to poor preventive maintenance, maloperation, or poor design. It has also been shown[10] that the incidence and repair times of failures can be considered as effectively random and the repair situations can therefore be represented by a queueing model. That is, for short periods the rate of failure exceeds the rate of repair and the items form a queue to wait for repair. In practice the queueing situation is complicated by the wide variety of corrective maintenance work with different repair priorities; this problem is discussed in detail in Chapters 5 and 6.

In *Figure 1.3* maintenance is presented as the operation of a pool of resources (men, spares and equipment) directed towards controlling the level of plant availability. The resources are divided between preventive maintenance (which might include design-out-maintenance modifications) and corrective maintenance. In order that the resources are directed in the best way it is necessary to establish a maintenance objective, a maintenance plan and a suitable maintenance organisation.

It was stated in Section 1.1 that the maintenance objective must be compatible with, indeed must be directed towards, company profitability. The objective is to achieve a balance between the costs of unavailability and the costs of the maintenance resources which exist to control unavailability. In the general model these costs are classified, for the sake of clarity, as follows:

Unavailability costs   Loss of in-service material. Production loss while in repair, waiting for repair or while undergoing preventive maintenance.

Resource costs   Corrective maintenance labour, preventive maintenance labour, maintenance equipment, spares usage and holding costs.

The proper maintenance objective in most industrial situations is the minimisation of the sum of the unavailability and resource costs. In many situations the cost of unavailability can vary greatly depending on sales and product storage factors. Where this is so the cost information must be available to the maintenance-production decision makers in the required detail and at the right time. The maintenance plan provides guidelines within which maintenance actions can be carried out. It can be seen from *Figure 1.3* that the plan should lead to the establishment of the preventive maintenance programme and should provide guidelines within which the corrective maintenance decisions can be taken.

The maintenance organisation establishes the level, mix, and distribution of resources, administrative structure, and work planning systems required to enable the maintenance load, corrective and preventive, to be dealt with in the most efficient manner.

Because of the complex and ever-changing nature of the maintenance problem it is not only necessary to establish a plan and an organisation but also to set up a control system to ensure that the plan and the organisation are continually updated. *Figure 1.4* shows that such a system performs three main inter-related functions: work control, plant condition control and cost control.

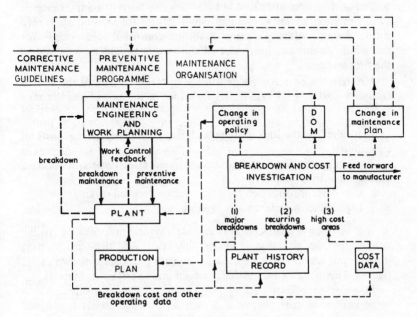

Figure 1.4. Maintenance control model

Work control is a function of the maintenance organisation and its object is to match men, spares and equipment to the maintenance work load (preventive and corrective). This function includes the location of plant failures, the determination of the necessary corrective action, the setting of priorities and the coordination and control of the men, spares and maintenance equipment (see Chapter 5).

Plant condition control is needed to achieve optimum plant performance in the long term and its function is to identify the most important problems, diagnose causes and prescribe solutions. In this respect (and taking into consideration the characteristics of equipment failure) the modification of preventive maintenance policy is only one of a number of alternative actions which would optimise plant performance. Others are equipment re-design (especially early in the equipment life) and changes in production policy. Because of the interdisciplinary nature of plant failure problems the more important failure analyses should be carried out by a small interdepartmental team. *An essential requirement for such an organisation is the efficient feedback, processing and analysis of failure and cost data*, much of which is also required for corrective maintenance decision making. An additional function of failure and cost investigations is the transmission of information to the equipment manufacturer for incorporation into future designs. Although plant condition control should operate for the whole equipment life, most of the results should be achieved in the first few years.

Maintenance cost control is usually operated as part of the overall company cost control system. Its functions in the case of maintenance should be

1. to identify the high cost areas of plant, a form of management by exception;
2. to monitor the trend of maintenance effectiveness, a form of management by objectives;
3. to provide information for maintenance decision making;
4. to facilitate maintenance budgeting.

It is instructive at this point to outline a typical maintenance costing system for the situation modelled in *Figure 1.3*; the basic costing information which might be used for control purposes is shown in the figure (in many cases unavailability would not be put in cost terms).

In order to collect costing information the plant is divided into areas known as *cost centres* and each unit of equipment is assigned to one of these. To facilitate identification and data processing a numerical equipment code, of the type shown below, is used.

Cost centre	Unit	Assembly	Sub-assembly
45	845	08	3

Figure 1.5 shows a costing system based on this type of code; it obtains inputs from three sources, namely

1. Downtime cards.
2. Time cards.
3. Stores requisition forms.

(An alternative, and more accurate, method of providing information on the time spent on jobs is through an engineering work order procedure of the type that will be discussed in Section 5.4.3.)

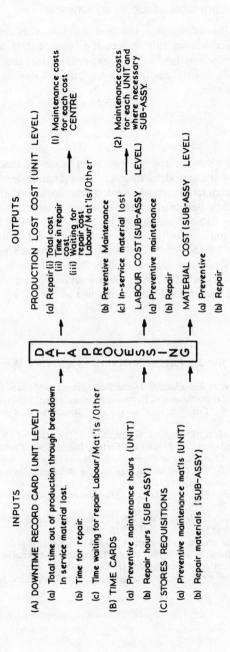

Figure 1.5. Costing system

INPUTS

(A) DOWNTIME RECORD CARD (UNIT LEVEL)

(a) Total time out of production through breakdown
 In service material lost.

(b) Time for repair:

(c) Time waiting for repair Labour/Mat'ls/Other

(B) TIME CARDS

(a) Preventive maintenance hours (UNIT)

(b) Repair hours (SUB-ASSY)

(C) STORES REQUISITIONS

(a) Preventive maintenance mat'ls (UNIT)

(b) Repair materials (SUB-ASSY)

DATA PROCESSING

OUTPUTS

PRODUCTION LOST COST (UNIT LEVEL)

(a) Repair (i) Total cost
 (ii) Time in repair cost.
 (iii) Waiting for repair cost
 Labour/Mat'ls/Other

(b) Preventive Maintenance

(c) In-service material lost

LABOUR COST (SUB-ASSY LEVEL)

(a) Preventive maintenance

(b) Repair

MATERIAL COST (SUB-ASSY LEVEL)

(a) Preventive

(b) Repair

(1) Maintenance costs for each cost CENTRE

(2) Maintenance costs for each UNIT and where necessary SUB-ASSY.

The nature of the information required will obviously depend on the situation, which will in turn govern the type of data collection and processing procedures used. In the example shown the situation was such that it was essential, both from the point of view of cost and time, to use a computer.

1.5 Summary

In this opening chapter an attempt has been made to describe the overall operation of the industrial maintenance system in terms of the causes, costs and characteristics of maintenance work; availability has been used as the link between maintenance and production. Such an analysis shows the close relationship between the production and maintenance functions, classifies the reasons for maintenance work, and delineates the interaction between the corrective and preventive maintenance functions. The general models discussed can be used for the analysis of most industrial maintenance situations and therefore provide a structural basis for the ideas in the rest of the book.

REFERENCES

1. Report by the Working Party on Maintenance Engineering, Department of Industry (1970)
2. *Terotechnology, an Introduction to the Management of Physical Resources*, Committee for Terotechnology, Department of Industry (1975)
3. Blanchard, S. B., *Logistics Engineering and Management*, Prentice Hall (1974)
4. Ablitt, J. F., Moss, T. R. and Westwell, F., 'The Role of Quantitative Assessment and Data in Predicting System Reliability', *Conference on Improvement of Reliability in Engineering*, I.M.E. (1974)
5. Blackman, W. D. H., 'Terotechnology in the R.E.M.E.,' *Iron and Steel Institute Conference*, May (1972)
6. Internal Report, Plant Availability Study Group, National Centre of Systems Reliability, UKAEA (1975)
7. Brook, R. H. W., *Reliability Concepts in Engineering Manufacture*, Butterworths (1972)
8. BS 3811, British Standards Institution (1974)
9. Steedman, J. B., and Treadgold, A. J., 'Collection and Analysis of Data on Chemical Plant', *Conference Pub. 11*, I.M.E. (1973)
10. Buffa, S. B., *Modern Production Management*, 757, Wiley (1965)

Chapter 2

Decision Making and Failure Statistics

2.1 The Nature of Decision Making

Maintenance management decisions, directed at every part of the maintenance function — its objectives, its organisational structure, its actions — inevitably cut across departmental boundaries and require information from diverse sources. Such decisions can range in importance from a question of major equipment replacement to a minor repair option. It is therefore important to understand the structure of the decision making process.

A decision problem exists when

1. there is a desired objective,
2. at least two courses of action are available,
3. there is uncertainty as to which course is the best,
4. external factors are present which can affect the outcome and which are outside the control of the decision maker.

An outline decision making procedure for such a situation is shown in *Figure 2.1*.

A typical maintenance problem is that presented by a recurrent equipment failure. The decision required is the selection of a course of action which will minimise its frequency, and the selection might be based, as it often is, on a minimum cost criterion. Possible courses of action might be as follows:

Redesigning to avoid failure.
Replacing at fixed time before failure.
Replacing on inspection before failure.
Replacing after failure.

13

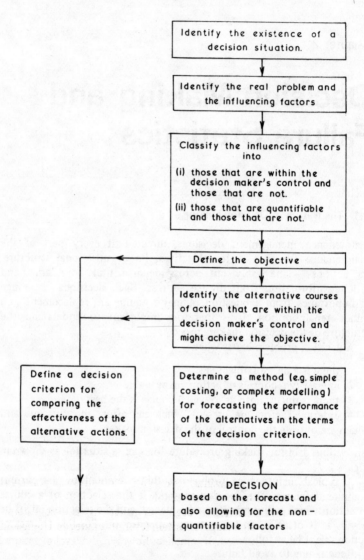

Figure 2.1. A decision making procedure

Identification of the most suitable action is only possible if there is understanding of the influencing factors, the most important of which would probably be the cause of failure, its incidence with respect to time, the costs of maintenance and the costs of re-design and the non-quantifiable elements such as the attitude of the trade force to preventive maintenance.

In some situations of the above kind a simple costing procedure and a little engineering judgement are all that are required to enable the alternatives to be ranked and the decision made. On the other hand, in order to forecast the performance of alternatives, it might be necessary to use complex models of the situation which involve statistical patterns of plant behaviour and sophisticated mechanisms of cost behaviour (the ultimate decision will, of course, still require judgement of the non-quantifiable factors discussed in detail in Chapter 4). A selection of techniques — e.g. queueing theory, simulation, statistical failure analysis — which can be incorporated in such models will be outlined in this book.

As stated at the start of this chapter, maintenance decisions vary greatly in importance. The higher level decisions are of the one-off type which might be irreversible and might involve large capital expenditure, e.g. a major equipment re-design. For these the procedure of *Figure 2.1* would be followed in detail and would thus necessitate substantial data collection and analysis. Equally important, but of a different type, is the decision situation in which a major item of plant presents a fairly frequent repair-*vs.*-replace problem[1]. The procedure in this case would be to provide objectives and a set of decision rules for the guidance of lower level decision makers, such as foremen; the complementary information systems must, of course, also be provided.

The last example illustrated the principle of ensuring that maintenance decision makers at all levels should be properly equipped for their task. It is not always possible, or indeed desirable, to provide decision rules, but it is absolutely essential to ensure that maintenance objectives are clearly understood at all levels and that the information systems have been designed to provide the right information to the right people at the right time.

2.2 An Introduction to Failure Statistics

Many of the problems in the field of maintenance and reliability are situations that involve probabilistic variables. The modelling of such situations therefore requires a basic understanding of failure statistics, i.e. of the application of statistical techniques to the description or analysis of patterns of failure of components and equipment. Before dealing with some detailed applications of failure statistics in maintenance and reliability some basic statistical ideas need to be examined.

2.2.1 Mean, Variance and Standard Deviation

One hundred filament lamps were run continuously, the time to failure noted in each case and *Table 2.1* drawn up.

Table 2.1 LAMP FAILURE DATA

Class interval	Frequency	Relative frequency	Rel. freq. density
Time to failure, h	No. of lamps	Fraction of total	Fraction per hour
300–400	2	0.02	0.0002
400–500	9	0.09	0.0009
500–	21	0.21	0.0021
600–	40	0.40	0.0040
700–	19	0.19	0.0019
800–	8	0.08	0.0008
900–	1	0.01	0.0001
	100	1.00	

Note that the figures in the fourth column are obtained by dividing those in the third by 100 h, the width of the class intervals used.

Using the data in the fourth column a *histogram* of the relative frequency densities can be constructed (see *Figure 2.2*). By drawing the histogram this way the *area* of the block above each class interval equals the *relative frequency* of failure in that interval (even if unequal class intervals are used, which is sometimes more convenient).

The assumption might now be made that the pattern of failure of this particular sample of 100 lamps represents the probable failure pattern of all such lamps, i.e. the observed relative frequencies truly reflect the expected probabilities of failure. The probability that a lamp of this kind will last longer than, say, 700 h is then given by the shaded area in this histogram, i.e. $0.19 + 0.08 + 0.01 = 0.28$.

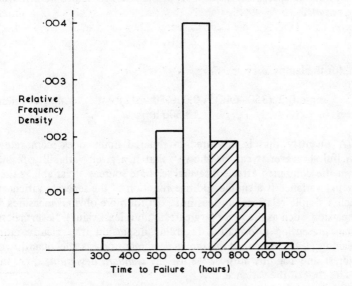

Figure 2.2. Histogram of data in Table 2.1

Having displayed the general form of the variability in time-to-failure some parameters are now required which can usefully characterise the various features of this variability.

1. For the typifying value, or *central tendency*, of the observed times-to-failure, we often use

$$\text{the } arithmetic \ mean, \ m = \Sigma \ f_r t_r$$

where

Σ = 'the sum of all such terms as . . . . .'

f_r = relative frequency in the rth class interval
t_r = mid point of the rth class interval.
Thus, for the sample of lamps,

$$m = (0.02 \times 350) + (0.09 \times 450) + (0.21 \times 550) + \ldots . \text{etc.}$$
$$= 643 \text{ h}$$

2. For the spread, or *dispersion*, of the times-to-failure we often use the *variance*,

$$s^2 = \Sigma f_r (t_r - m)^2$$

So, for the lamps

$$s^2 = 0.02(350-642)^2 + 0.09(450-642)^2 + \ldots\ldots \text{etc.}$$
$$= 13\,500\,\text{h}^2$$

A quantity that is measured in squared hours does seem rather fanciful; however it can be shown[2] that if a given variability results from the combined effect of several separate sources of variability then the net variance is a simple additive function of the separate variances. Such a simple relationship does not exist for more obvious measures of dispersion, such as the *range* (= greatest minus least value). Nevertheless, when presenting information regarding dispersion it is clearly more meaningful to present it in the same dimensions as the quantity of interest and for this purpose we use the *standard deviation, s*, i.e. the square-root of the variance.

So, for the lamps,

$$s = (13\,500)^{1/2} = 116\,\text{h}$$

2.2.2 Probability Density Functions

If, instead of just 100 lamps, many thousands had been tested the width of the class intervals in *Figure 2.2* could have been reduced to such a degree that the step nature of the histogram would have been virtually eliminated and a continuous probability density distribution obtained, as in *Figure 2.3*. Many failure-causing mechanisms give rise to measured distributions of times-to-failure which approximate quite closely to probability density distributions of definite mathematical form, known as *probability density functions*, or p.d.f.s. Such functions therefore provide mathematical models of failure patterns, which can be used in performance forecasting calculations.

The negative exponential p.d.f.   Experience with a very wide range of components and equipment shows that, under normal operating conditions and during their normal operating life, they do not reach a point of wear-out failure at some likely time that could be called 'old-age'. On

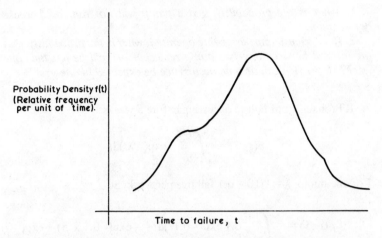

Figure 2.3. Continuous probability density distribution

the contrary, a given item is as likely to fail in a given week shortly after installation as in a given week many months later. In short, *the probability of failure is constant and independent of running time; the item is always effectively 'as good as new'.* Very often, such behaviour indicates that the cause of failure is external to the item. A fuse is always as good as new until a short circuit elsewhere in the system blows it. Whitaker[3] quotes a case in the chemical process industry where age-independent failure of pump seals was caused by gas locking and heat checking due to inappropriate design of other parts of the process flow path.

If this is the case, then it can be shown (see Chapter 4 of Ref. 2) that the p.d.f. of time, *t*, to failure is given by the expression

$$f(t) = \lambda \exp(-\lambda t)$$

where
λ = average failure rate (failures/unit time) per machine. (It follows that $1/\lambda$ = average time to failure.) This is known as the *negative exponential p.d.f.*, or more briefly the *exponential p.d.f.*

Example. Under given operating conditions a particular type of pump shows an average time to failure of 10 weeks. On failing it is replaced and overhauled to an as-good-as-new condition, but the probability of failure is found to be independent of time from replacement.

1. What is the probability that a pump will not run for 5 weeks before failing?

2. If 10 such pumps are being operated, what is the probability that the interval between any two pump replacements will be less than one week? How often can such an occurrence be expected in one year?

1. Probability of failure occurring before 5 weeks,

$$p(t<5) = \int_{t=0}^{t=5} \lambda\exp(-\lambda t)\mathrm{d}t$$

For one pump, $\lambda = 1/10 = 0.1$ failures per week. So

$$p(t<5) = \int_{t=0}^{t=5} 0.1\exp(-0.1t)\mathrm{d}t = -\exp(-0.1 \times 5) + \exp(-0)$$

$$\cong -0.61 + 1 = 0.39$$

2. For 10 pumps, $\lambda = 1$ failure per week. So

$$p(t<1) = \int_{t=0}^{t=1} \lambda\exp(-\lambda t)\mathrm{d}t = \int_{t=0}^{t=1} 1\exp(-1t)\mathrm{d}t = -\exp(-1) + 1$$

$$= 0.63$$

Average failure rate = 1 per week, so expected number of between-failure intervals in one year = 52.

Therefore, expected number of intervals of less than one week

$$= 0.63 \times 52 \cong 32$$

The hyper-exponential or 'running-in' p.d.f. With many types of equipment the probability of failure is found to be much higher during the period following installation than during its subsequent useful life. Such behaviour results in a p.d.f. of time-to-failure which, by contrast with the negative exponential p.d.f. which shows a single exponential fall-off, exhibits two phases — an initial rapid exponential fall and a later slower exponential fall. This is illustrated in Figure 2.4. It is evidence that some of the items are manufactured or installed with

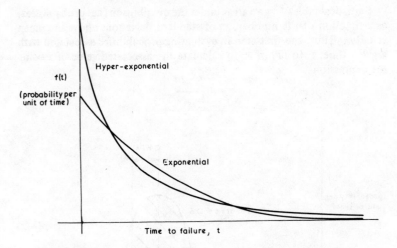

Figure 2.4. Comparison of hyper-exponential and exponential p.d.f.s

built-in defects which show up during the running-in stage. Those that survive this stage without failure were without such defects to begin with, they go on to exhibit the sort of time-independent failure probability previously discussed.

Note that the equipment is not improving with age! Some items merely start off with a better chance or survival than others.

The normal or 'wear-out' p.d.f. Many items, e.g. filament lamps as in Section 2.2.1, do show a marked wear-out failure pattern. They tend to fail at some mean operating age, m, with some failing sooner and some later, thus giving a dispersion, of standard deviation s, in the recorded times to failure. The p.d.f. of time-to-failure often approximates quite closely to the expression

$$f(t) = \frac{1}{s\sqrt{(2\pi)}} \exp\left\{-\frac{(t-m)^2}{2s^2}\right\}$$

which is known as the *normal* p.d.f., a symmetric bell-shaped curve as shown in *Figure 2.5*. As indicated, 50% of all items would be expected to show times to failure in the range $(m - 0.67s)$ to $(m + 0.67s)$, 95% in the range $(m - 2s)$ to $(m + 2s)$.

Statistical tables [4,5] give areas under the distribution (i.e. probabilities) as a function of the number, x, of standard deviations above the mean value, m. Thus, the first step in evaluating probabilities associated with a given time, t, to failure is to calculate the associated value of x using the expression

$$x = \frac{t - m}{s}$$

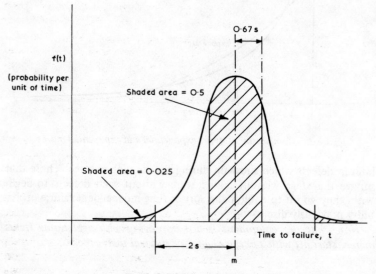

Figure 2.5. The normal p.d.f.

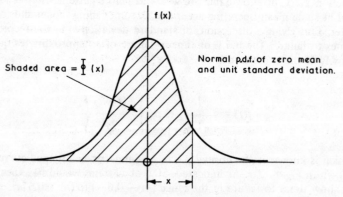

Figure 2.6. The probability Φ (x)

Table I, Ref.4, gives the probability, $\Phi\,(x)$, of occurrence of a value less than x (see *Figure 2.6*), i.e. the probability of failure occurring before the corresponding time t. Most compilations of statistical tables include a table of this type.

Example. Filament lamps are found to show times to failure which are normally distributed with a mean of 120 weeks and a standard deviation of 20 weeks. What proportion may be expected to last
1. *150 weeks or more?*
2. *More than 70 and less than 150 weeks?*

$$1. \qquad x = \frac{t - m}{s} = \frac{150 - 120}{20} = 1.5$$

From Table I, Ref. 4, $\Phi\,(1.5) = 0.9332$ = probability of failure *before* 150 weeks. So, probability of failure after 150 weeks = $1 - 0.9332 = 0.0668 = 6.68\%$

$$2. \qquad t = 70, x = \frac{70 - 120}{20} = -2.5$$

$$t = 150, x = \frac{150 - 120}{20} = +1.5$$

It is now necessary to evaluate the shaded area of *Figure 2.7.*

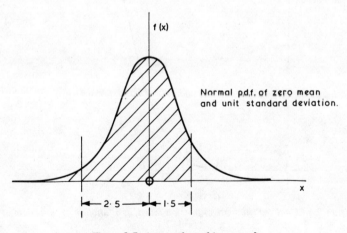

Figure 2.7. Area evaluated in example

As before, from Table I. Ref. 4,

Probability of $x > 1.5 = 1 - \Phi(1.5) = 1 - 0.9332 = 0.0668$

The normal p.d.f. is symmetric, so

Probability of $x < -2.5$ = probability of $x > 2.5 = 1 - \Phi(2.5)$
$$= 1 - 0.9938 = 0.0062$$

So, probability of $1.5 > x > -2.5 = 1 - (0.0668 + 0.0062)$
$$= 0.927 = 92.7\%$$

2.2.3 *Failure Probability, Survival Probability, Age-Specific Failure Rate*

The time-to-failure information in *Figure 2.2* can be presented in other forms that may be more useful for the study of reliability.

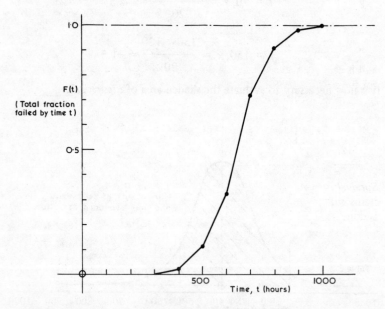

Figure 2.8. Total (or cumulative) fraction failed, as a function of time; data of Table 2.1

Failure probability A graph can be drawn (see *Table 2.2* and *Figure 2.8*) of the rise in the total fraction, $F(t)$, of the items that have failed by any given time t.

Table 2.2 VARIATION WITH TIME OF TOTAL FRACTION FAILED: DATA OF TABLE 2.1

Time	t, h	0	100	200	300	400	500	600	700	800	900	1000
Total fraction failed	$F(t)$	0	0	0	0	0.02	0.11	0.32	0.72	0.91	0.99	1.00

If the sampled 100 lamps are assumed to be representative then $F(t)$ is the fraction of all such lamps which can be expected to fail by the running time, t, since new. That is, for any given lamp, $F(t)$ = probability of failure before a running time t.

For one of the analytical p.d.f.s of Section 2.2.2

$$F(t) = \int_0^t f(t)\mathrm{d}t$$

and is called the *cumulative distribution function* or c.d.f. For example, for the negative exponential p.d.f.

$$F(t) = \int_0^t \lambda \exp(-\lambda t)\mathrm{d}t = 1 - \exp(-\lambda t)$$

Survival probability By contrast with the above the fraction, $P(t)$, of items *surviving* at running time t could be tabulated and plotted. This is done in *Table 2.3* and *Figure 2.9*.

Table 2.3. VARIATION OF FRACTION SURVIVING; DATA OF TABLE 2.1

Time	t, h	100	200	300	400	500	600	700	800	900	1000
Fraction	$P(t)$	1.0	1.0	1.0	0.98	0.89	0.68	0.28	0.09	0.01	0.00

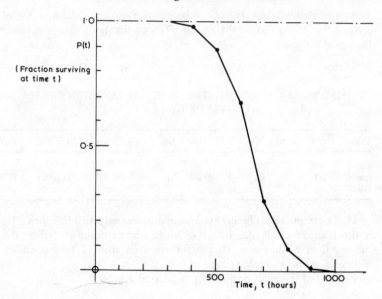

Figure 2.9. Fraction surviving, as a function of time; data of Table 2.1

Clearly, $P(t) = 1 - F(t)$, and for the negative exponential case, for example,

$$P(t) = \exp(-\lambda t)$$

Again, assuming that the 100 lamps are representative, $P(t)$ is the *survival probability*, at time t, for any one lamp. In the literature on this subject $P(t)$ is very commonly called the 'reliability', at time, t, and the symbol $R(t)$ is used. The term survival probability is used here in order to distinguish this measure of reliability from that which is appropriate in the case of things like pressure release valves, temperature sensitive trip-mechanisms, and safety mechanisms generally, which are normally inoperative until called on. Expected fraction of demands meeting with the required response would be a more relevant parameter of reliability in that context.

Age-specific failure rate   This is defined as the fraction, $Z(t)$, of those items which have survived up to the time, t, which can be expected to fail in the next unit of time. This is calculated, tabulated and plotted, for the lamps of *Table 2.1*, in *Table 2.4* and *Figure 2.10*.

Table 2.4. VARIATION OF AGE-SPECIFIC FAILURE RATE;
DATA OF TABLE 2.1

Time t, h	100	200	300	400	500	600	700	800	900	1000
Fraction failing in $f(t)$ interval after t	0	0	0.02	0.09	0.21	0.40	0.19	0.08	0.01	0
Fraction surviving $P(t)$ at time t	1	1	1	0.98	0.89	0.68	0.28	0.09	0.01	0
Age-specific failure rate $Z(t) = \dfrac{f(t)}{P(t)}$	0	0	0.02	0.09	0.24	0.59	0.68	0.89	1.00	

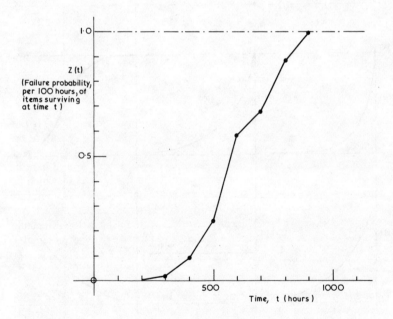

Figure 2.10. Age specific failure rate as a function of time; data of Table 2.1

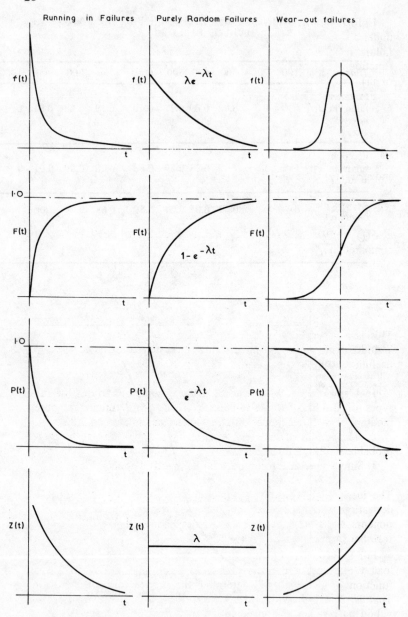

Figure 2.11. Comparison of statistical functions for principal modes of failure

Unfortunately, in the literature on the subject, this quantity goes under a variety of titles such as 'age-specific failure rate', 'instantaneous failure rate', 'local failure rate', 'local probability of failure', 'hazard function', or simply 'failure rate' (which is the most loose terminology of all).

For an analytical p.d.f.,

$$Z(t) = \frac{f(t)}{P(t)} = \frac{f(t)}{1 - F(t)}$$

and again, for the negative exponential case,

$$Z(t) = \lambda\exp(-\lambda t)/\exp(-\lambda t) = \lambda$$

$F(t)$, $P(t)$, and $Z(t)$ for the p.d.f.s discussed in Section 2.2.2 are compared in *Figure 2.11*.

2.2.4 The Weibull p.d.f.

This is a particularly useful semi-empirical expression[6] developed by Waloddi Weibull for use in his studies of the strengths of steels. Its usefulness in the context of this book lies in its providing:

1. A single p.d.f. which can be made to represent any of the three types of p.d.f.s of times-to-failure (arising from running-in, purely random, or wear-out modes of failure) described in Section 2.2.2.
2. Meaningful parameters of the failure pattern, such as the probable minimum time to failure.
3. Simple graphical techniques for its practical application.

The ideas underlying this p.d.f. may be grasped from Weibull's own derivation, which pursued the analogy between the crystalline components which together form a steel specimen and the links which together form a chain.

For individual chain links the probability of failure, $F(x)$, under a load x must be of the form shown in *Figure 2.12* (where $\phi(x)$, some function of the load, will determine the precise form of the curve). That is, if a thousand separate links were tested, the fraction that would have failed at a given load would increase with the size of the load and would approach unity at high loads.

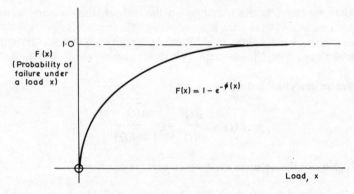

Figure 2.12. Probability of failure of a chain link as a function of load

So, the probability of survival of a given link under a load x would be

$$P(x) = 1 - F(x) = \exp(-\phi(x))$$

and the probability of survival of a chain of n links would be

$$P_n(x) = [\exp\{-\phi(x)\}]^n = \exp(-n\phi(x))$$

and the probability of failure

$$F_n(x) = 1 - \exp(-n\phi(x))$$

which is the appropriate general expression for the failure probability of a chain, where total failure arises from failure of the weakest link and therefore where the likelihood of failure increases with the number, n, of links. Weibull used it to correlate failure data for test specimens of different sizes (on the general analogy between size of the specimen, and hence number of crystalline components, and number of links in a chain).

A definite form is still needed for the function $\phi(x)$. This should be such that

1. $F(x)$ should never decrease as x increases.
2. $F(x)$ should be zero at some threshold load x_0.
3. $\phi(x)$ should be dimensionless (since it is an exponent).

Weibull suggested the form

$$\phi(x) = \left\{\frac{x - x_0}{\eta}\right\}^\beta$$

where

x_0 = threshold load
η = characteristic load
β = shape factor.

This gives the Weibull c.d.f.

$$F(x) = 1 - \exp\left\{-\left(\frac{x - x_0}{\eta}\right)^\beta\right\}$$

In the reliability problems discussed in this chapter the stressing agent is not load but running time, t, since new or since last overhaul. The Weibull c.d.f. is then written as

$$F(t) = 1 - \exp\left\{-\left(\frac{t - t_0}{\eta}\right)^\beta\right\}$$

Straightforward mathematics based on the relationships in Section 2.2.3 then leads to the other probability functions, i.e.

$$P(t) = \exp\left\{-\left(\frac{t - t_0}{\eta}\right)^\beta\right\}$$

$$f(t) = \frac{\beta(t - t_0)^{\beta-1}}{\eta^\beta} \exp\left\{-\left(\frac{t - t_0}{\eta}\right)^\beta\right\}$$

and

$$Z(t) = \frac{\beta}{\eta^\beta}(t - t_0)^{\beta-1}$$

Each of the terms t_0, η and β has a very practical significance.

The threshold time-to-failure, or guaranteed life, t_o. In many cases of wear-out the first failures do not appear until some significant running time t_o has elapsed. Age specific failure rate $Z(t)$ is non-zero and rising only after t_o, so in the Weibull expressions the time factor always occurs as the time interval $(t-t_o)$.

The characteristic life, η. When $t-t_o = \eta$, $P(t) = \exp(-1) = 0.37$, i.e. η is the interval between t_o and the time at which it can be expected that 63% of the items will have failed and 37% survived.

The shape factor, β. Figures 2.13 and *2.14* show how the various patterns of time-to-failure and age-specific failure rate are characterised by the value of β. A 'running-in' or 'infant-mortality' failure pattern leads to a value significantly less than one, a negative exponential (random failure) pattern to a value fairly close to one, and a wear-out pattern to larger values, although if β is less than, say, 3 then a purely random factor is still significant.

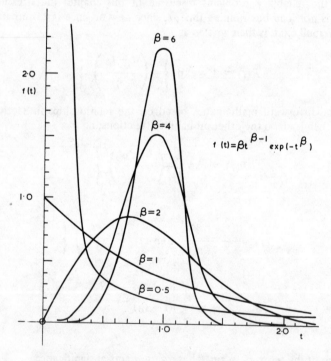

Figure 2.13. The Weibull p.d.f. (for simplicity, $\eta = 1$, $t_o = 0$)

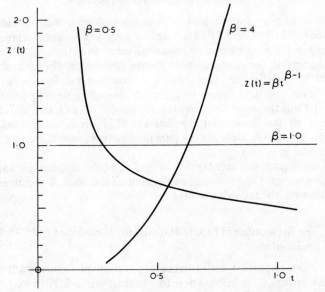

Figure 2.14. Weibull age specific failure rate curves (for simplicity, $\eta = 1$, $t_o = 0$)

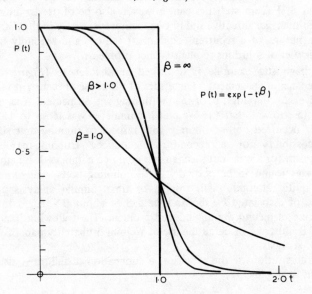

Figure 2.15. Weibull survival probability curves (for simplicity, $\eta = 1$, $t_o = 0$)

It is illuminating to look at the survival probability pattern as illustrated in *Figure 2.15*. The larger the value of β the greater is the tendency for all the items to fail at the same time, $(t_0 + \eta)$. Thus, β is a measure of the consistency of failure occurrence. (In passing, it is worth noting that the infinite β case illustrates the difference between two often confused parameters, namely the mean time to failure (or MTTF) and the mean time between failures (or MTBF). In this extreme case if all the items start together the MTTF is of value η and the MTBF of value 0. Only for the pure negative exponential case are these two equal in value.)

Simple graphical techniques, using a specially designed probability graph paper, for the practical application of the Weibull function, will be shown in the following section.

2.3 The Application of Failure Statistics to Maintenance and Reliability Engineering

The most important maintenance application of failure statistics is in the provision of information to designers and reliability engineers, enabling them to determine system reliabilities, availabilities, expected lives, etc., with greater certainty. This aspect will be discussed in Chapter 3. Failure statistics can, however, also be of use to the maintenance manager directly, and in two main ways, firstly in the *diagnosis* of the nature of a recurrent equipment failure, and secondly in the *prescription* of solutions to maintenance problems.

A diagnostic example is described in Chapter 13 (*Figure 13.5*). Maintenance management were unsure of the cause of recurrent failures of universal couplings; poor reconditioning was suspected. Analysis of the data showed that the Weibull β parameter was close to 1.0, i.e. failures occurred quite randomly with respect to running time so that the possibility of a reconditioning-induced (running-in) failure mechanism (or a wear-out mechanism) could be eliminated. Finally, the cause was found to be poor design of coupling bolts, which worked loose quite randomly with respect to time. Similar analysis of the failures of associated gearboxes gave a β of about 0.5, suggesting, in their case, a reconditioning-induced problem. The following Examples A and B illustrate in detail the use of Weibull probability paper for this type of analysis.

An illustration of the prescriptive application of failure statistics is given in Example C.

Example A. Weibull analysis of a large and complete sample of times-to-failure.

One hundred identical pumps have been run continuously and their times-to-failure noted. A Weibull fit to the data is required. The procedure is as follows.

1. The date is tabulated as in columns 1 and 2 of *Table 2.5*.

Table 2.5 PUMP FAILURE DATA

1	2	3	4	5	6
Time-to-failure, h	Number of pumps	Cumulative per cent failed	$t-t_0$ (t_0=800 h)	$t-t_0$ (t_0=900 h)	$t-t_0$ (t_0=1000 h)
1000–1100	2	2	300	200	100
1100–1200	6	8	400	300	200
1200–1300	16	24	500	400	300
1300–1400	14	38	600	500	400
1400–1500	26	64	700	600	500
1500–1600	22	86	800	700	600
1600–1700	7	93	900	800	700
1700–1800	6	99	1000	900	800
1800–1900	1	100	1100	1000	900

2. Successive addition of the figures in column 2 leads to column 3, the total percentages of pumps failed by the times, t, that each of the intervals in column 1 came to an end.

3. Three or four possible values, thought likely to span the actual value, are assigned to t_0 (the threshold time-to-failure). The resultant values of $t-t_0$ are tabulated in columns 4,5,6, etc.

4. Weibull probability paper (in this case Chartwell Graph ref. 6572) is used to plot the column 3 figures against those in columns 4,5 and 6 respectively. The result is shown in *Figure 2.16*, which also includes plots with t_0 = 700 h and t_0 = 1100 h. The value of t_0 finally adopted (in this case 900 h) is that which results in the straightest plot.

5. The characteristic life, η, is the value of $t-t_0$ (in this case 600 h) at which the line fitted to the straightest plot reaches the 63% failed level. (Note that $t - t_0$ = 600 h corresponds to a total actual running time of t = 1500 h, remembering that t_0 = 900 h.)

6. As shown, a perpendicular is dropped from the *estimation point* to the straight line fit. The point at which this intersects the special scale at the top of the graph gives the value of β (in this case 3.5 approximately). Note that the perpendicular also intersects another scale which indicates the value of the cumulative per cent failed, F, at

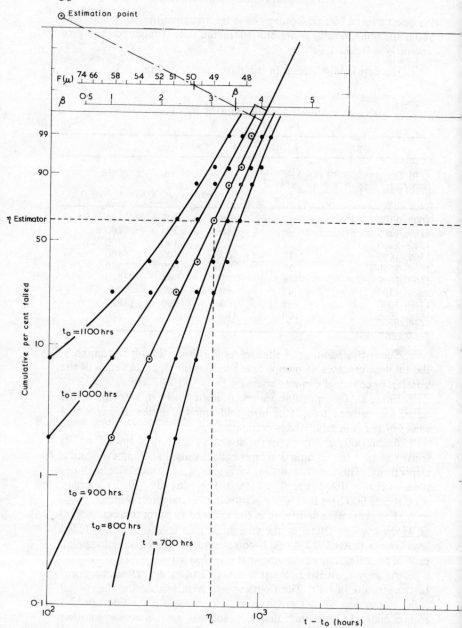

Figure 2.16. Weibull plot of pump failure data (from Table 2.5)

the point $\bar{t} - t_0$. In this case $F = 49.8\%$, which corresponds to $\bar{t} - t_0 = 540$ h , or $\bar{t}$ (the mean pump life) = 1440 h.

To sum up, the observations fit a Weibull p.d.f. of parameters

$t_0 = 900$ h,
$\eta = 600$ h,
and $\beta = 3.5$.

We have also deduced that

$$\bar{t} = 1440 \text{ h}$$

Example B. Weibull analysis of a small and incomplete sample.

In the previous example sufficient data had been obtained to enable direct calculation of F for a range of values of t. The values calculated were also likely to be not too different from those which would have been obtained had enough time or money been available to have measured several thousand times to failure.

In practice, there may only have been opportunity to measure a handful of times to failure. Indeed, the items under examination might be large, expensive and of low failure rate, and as is likely in such a case, only a few might yet have been made. In addition, some of them might still be running, not having reached the failure point (i.e. 'suspended'), or some of the tests may have been terminated (i.e. 'censored') before failure because, in their case, the test conditions were accidentally altered. In this situation the results of any analysis will necessarily be subject to greater statistical uncertainty, but a Weibull analysis may be required, on the grounds, for example, that an approximate result at the end of a fortnight may be of more value than a precise one obtained by waiting for another three months. A technique using *median ranks*, as demonstrated in the following example, is then appropriate. The statistical reasoning underlying the method is fairly sophisticated and for a more full explanation, illustrated by practical examples, the reader should consult Ref. 7.

Table 2.6.

Spring number	Cycles to failure	Spring number	Cycles to failure
1	9 100	6	(7200)*
2	8 000	7	4500
3	6 300	8	(5000)†
4	11 100	9	8400
5	3 300	10	5200

* Still running.
† Test terminated, inadvertent overspeed.

Ten oscillating springs are being tested to failure. The situation to date is as shown in *Table 2.6*. A Weibull fit to the data is required. The procedure is as below.

1. The failure points are ranked in ascending order (column 2 of *Table 2.7*) and classified as failed, f, or suspended (or censored), s, (column 3).

Table 2.7. SPRING-FAILURE DATA

1	2	3	4	5	6
Spring number	Cycles	Class	New increment	Order number	Median rank
5	3 300	f	1	1	0.067
7	4 500	f	1	2	0.163
8	5 000	s	–	–	–
10	5 200	f	1.125	3.125	0.272
3	6 300	f	1.125	4.250	0.380
6	7 200	s	–	–	–
2	8 000	f	1.350	5.600	0.510
9	8 400	f	1.350	6.950	0.639
1	9 100	f	1.350	8.300	0.770
4	11 100	f	1.350	9.650	0.898

2. For the first failed item the *new increment* is calculated from the formula

$$\text{New increment} = \frac{N + 1 - (\text{Order number of previous failed item})}{N + 1 - (\text{Number of previous items})}$$

where N = total number of items in the sample (i.e. 10). Since this is the first failed item, the previous order number is zero. Also, in this case, the number of previous items is zero and therefore the calculated new increment is 1 (column 4). Note that if the first failure had been preceded by some suspended items the new increment would have been greater than 1, e.g. if the first two items had been suspended, the new increment would have been $(10+1-0)/(10+1-2) = 1.22$.

3. The order number of the first failed item is obtained from the expression

Order number = New increment + Previous order number

i.e. in this case, order number = 1 + 0 = 1 (column 5).

4. This procedure is repeated for all the remaining *failed* items, in succession, i.e.

$$\text{Second failed item : new increment} = \frac{10 + 1 - 1}{10 + 1 - 1} = 1$$

$$\text{Order number} = 1 + 1 = 2$$

$$\text{Third failed item : new increment} = \frac{10 + 1 - 2}{10 + 1 - 3} = 1.125$$

$$\text{Order number} = 1.125 + 2 = 3.125$$

etc., etc.

The value of the new increment obtained for the first failed item after a suspended item remains constant, and therefore need not be revised, until the next group of suspended items.

5. Having completed column 5, the corresponding median ranks (column 6) are calculated from the formula

$$\text{Median rank} = \frac{\text{Order number} - 0.3}{N + 0.4}$$

e.g. for the fourth failed item

$$\text{Median rank} = \frac{4.250 - 0.3}{10 + 0.4} = \frac{3.950}{10.4} = 0.380$$

6. The median ranks, expressed as percentages, are plotted against cycles (or time) to failure on Weibull graph paper, as in *Figure 2.17*, and values of η, β and μ (8200 cycles, 2.78, and 7400 cycles respectively) obtained, as in Example A. In this case $t_0 = 0$, this giving a good straight line plot, but in the general case t_0 would be established by trial plots exactly as previously.

7. Johnson[7] gives tables of 5% and 95% ranks, for various sample sizes and order numbers. It is a simple matter to plot these, as well as the median (or 50%) ranks, against the observed cycles (or times) to failure, thus obtaining (see *Figure 2.17*) the 90% confidence band for the data, i.e. the band within which it is 90% probable that the plot *obtained from a very large number of items* would lie.

In the absence of suspended items the procedure is simpler, in that the order numbers are simply $1, 2, 3, 4 \ldots \ldots \ldots N$.

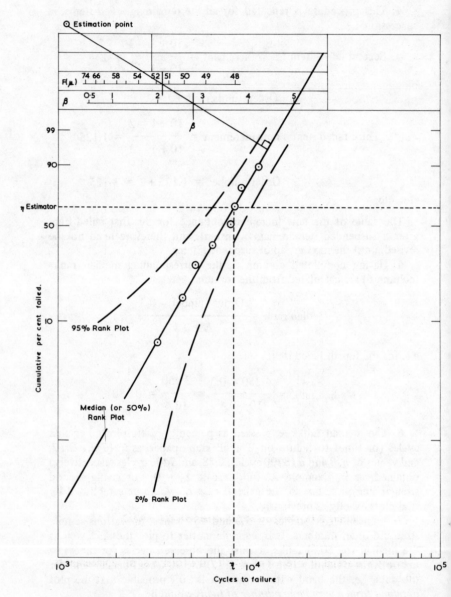

Figure 2.17. Weibull plot of spring failure median ranks (data of Table 2.6; *5% and 95% ranks also shown)*

Example C. Estimation of optimum fixed time replacement period for a wear-out item.

A certain type of component exhibits a wear-out pattern of failure as shown in *Figure 2.18.* When one fails it is replaced promptly at a cost C_f, of £3 per item. At periodic intervals, of length t, all such components are replaced at a cost, C_s, of £1 per item. The total number, N, of these components is 1000.

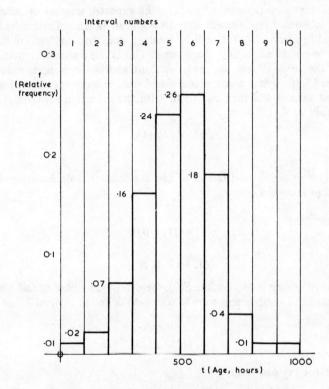

Figure 2.18. Component failure pattern, Example C

In any replacement interval of length t the total cost of replacements will be

$$NC_s + E(t)C_f$$

where $E(t)$ is the expected number of failure replacements during the interval t. The average total replacement cost per unit time will therefore be

$$C = \frac{NC_s + E(t)\,C_f}{t}$$

and the objective is to find that value of the replacement period which will minimise C. The rather lengthy part of this is the calculation of $E(t)$.

The first step is to calculate $E(1)$, the expected number of failures if the scheduled replacement were to occur at the end of interval 1 (as defined in *Figure 2.18*). It will be assumed that any lamp can fail only once, or not at all, in any one interval. This is not a serious restriction since the intervals can be made as short and hence as numerous as required (although the arithmetic would then become even more long-winded than it will turn out to be with the present ten intervals). It then follows that

$$E(1) = Np(1),$$

where

$p(1)$ = probability of failure of any one lamp if replacement occurs at end of interval 1.

In this case,

$$p(1) = f(1) = 0.01$$

so

$$E(1) = 0.01\,N.$$

The next step is to calculate $E(2)$, the expected number of failures if the scheduled replacement were to occur at the end of interval 2.

Now,

$$E(2) = N\,p(2)$$

where, for any one lamp,

$p(2) =$ (Probable number of failures in intervals 1 and 2 if first failure occurs in interval 1)
X (Probability of first failure occurring in interval 1)

PLUS
(Probable number of failures in intervals 1 and 2 if first failure occurs in interval 2)
X (Probability of first failure occurring in interval 2)

i.e.

$$p(2) = (1+p(1)).f(1)+1.f(2)$$
$$= 1.01 \times 0.01 + 1 \times 0.02 = 0.030$$

so

$$E(2) = 0.030\,N$$

Carrying this process another interval further,

$$p(3) = (1 + p(2)).f(1) + (1 + p(1)).f(2) + 1.f(3)$$
$$= 0.01 \qquad\quad + 0.02 \qquad\qquad + 0.07 = 0.100$$

so

$$E(3) = 0.100\,N.$$

Further extension leads, successively, to $E(4)$, $E(5)$, etc. Note that, in general

$$p(n) = \sum_{j=n-1}^{j=0} (1 + p(j))f(n-j)$$

which is a *recurrence formula*. To evaluate it for a given n it must have been evaluated for all smaller values of n.

Having calculated values of $E(t)$ it is now possible to find that value of t which minimises C. C is plotted, as a function of t, in *Figure 2.19*, from which it can be seen that the optimum replacement period is approximately 300 h.

Comparable models have been developed for other policies, e.g. optimum age replacement, group replacement, opportunistic replacement, etc. In practice, component replacement studies indicate that, more often than not, failure replacement is the cheapest solution. Fixed-time replacement is cheaper only if

1. the cost per item of failure replacement is much greater than that of fixed time replacement;

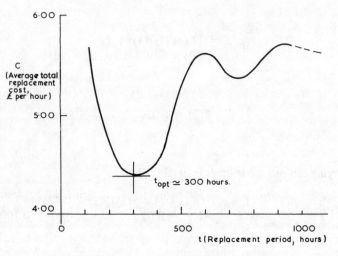

Figure 2.19. Cost vs. replacement period, Example C

2. the item concerned exhibits a pronounced wear-out failure behaviour (Weibull $\beta > 3$, say).

2.4 The Limitation of Failure Statistics in Maintenance Management

Learning the principles and techniques of statistical failure analysis, reliability design or replacement theory is one thing: collecting and making readily available a sufficient body of dependable data for these purposes is quite another. It cannot be done without adopting pains-taking, tightly controlled, and expensive procedures. Broadly speaking, the difficulties involved are of two sorts, arising from human factors on the one hand and equipment factors on the other.

With regard to human factors it is clear that great resistance exists at all levels (especially at shop floor level, expressed via its representative trade unions) to the collection of failure data. Many data collection systems studied by the authors have been less than successful because of shop floor resistance or apathy and lack of commitment by the immediate management. Often, this was no surprise because the system had been badly designed, and foisted on to the maintenance department without adequate prior consultation.

Some common faults were

1. Insufficient consideration of the motives for data collection, much of which was therefore unnecessary and manifestly so.

2. Insufficient appreciation of the problems of analysing the data so as to provide decision making information to the right people at the right time.

3. Over-elaboration and excessive demands on the data collectors.

When designing a data collection system (see Chapter 5) it is vital to aim at the maximum possible simplicity. A system providing limited, but correct, information at the right time is infinitely preferable to one which is sophisticated but unreliable and installed merely for managerial window dressing.

Equipment factors can also prove a serious obstacle. Clearly, data can be accumulated more rapidly, and its subsequent analysis be more fruitful, in those industries (transport, the armed services, communications, forging, etc.) in which many identical items, in comparable environments, can be observed concurrently. If only a single item exists, or is available for observation, times-to-failure have to be observed between consecutive replacements or repairs and by the time sufficient data is accumulated its analysis may be of diminished value. Very often the cause of recurring failures can be diagnosed directly and designed out, thus terminating the original study.

Probably the leading U.K. organisation for reliability data collection and analysis is the National Centre of Systems Reliability (N.C.S.R.) based at Culcheth, Lancs. Its activities are an external development of the work of the Safety and Reliability Directorate of the U.K. Atomic Energy Authority. A data collection system, a computerised reliability data bank (SYREL), and a reliability consultancy service are operated for what is effectively a world-wide club of diverse companies and national organisations.

REFERENCES

1. Armitage, W., *Maintenance Effectiveness. Operational Research in Maintenance,* (Jardine, A., Ed.), M.U.P. (1970)
2. Paradine, C. G. and Rivett, B. H. P., *Statistical Methods for Technologists,* E.U.P. (1953)
3. Whitaker, G. D., 'Statistical Reliability Models for Chemical Process Plant', *Symposium on Design for Reliability*, Inst. Chem. Eng., Apr. (1973)
4. Lindley, D. V. and Miller, J. C. P., *Cambridge Elementary Statistical Tables,* C.U.P. (1966)
5. Murdoch, J. and Barnes, J., *Statistical Tables*, MacMillan (1970)
6. Weibull, W., 'A Statistical Distribution Function of Wide Applicability', *Journal of Applied Mechanics,* 293, Sept. (1951)
7. Johnson, L. G., *Theory and Technique of Variation Research*, Elsevier (1964)

Chapter 3

Reliability Engineering and Maintenance

3.1 Introduction

Reliability theory is a young subject. It seems that the first major developments in this field were concerned with military aircraft (where the advantages of high reliability are self-evident) and occurred in the nineteen forties. Lomnicki[1] tells of a colleague who, preparing a paper on air safety, found that the library of the Institution of Mechanical Engineers had neither paper nor textbook on reliability; this was in March 1949. The first textbook in the English language appears to be that of Bazovsky[2]. It was published in 1961.

In the last few years pressure to obtain economies of scale has resulted in an unparalleled growth in unit sizes of equipment in most industries (e.g. oil and petro-chemical processing plant, passenger aircraft, electricity generating sets) with the result that the consequence of failure has become either much more expensive, as in the case of low availability of a large power station, or potentially catastrophic, as in the failure of a nuclear reactor shut-down mechanism. It is therefore becoming more and more important to be able to predict the expected life of plant and its major parts, the availability of plant, the expected maintenance load, and hence the support system resources needed for effective operation. Such prediction can only result from careful consideration of reliability and maintainability factors at the design stage. It is this relationship, between reliability design and consequent maintenance load and strategy, that renders some treatment of reliability engineering obligatory in a book of this kind.

46

3.2 The Whole-life Equipment Failure Profile

By combining the three $Z(t)$ curves of *Figure 2.11* a single $Z(t)$ curve as in *Figure 3.1* is obtained which, very broadly speaking, gives the whole-life profile of failure probability for the generality of components. The absolute levels of $Z(t)$, the time scale involved, the relative lengths of Phases I, II and III, will vary by orders of magnitude from one sort of item to another — indeed one or two of the phases could be effectively absent (e.g. high reliability aircraft control gear, where I

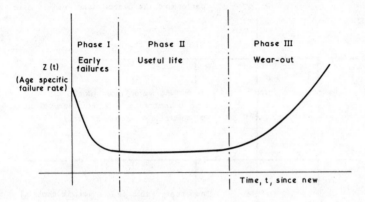

Figure 3.1. Whole life Z(t) curve (the 'Bath Tub' curve)

is negligible and III non-existent) — but the general behaviour will be as shown. It is interesting to note that human beings strikingly exemplify this behaviour. Greene and Bourne[3] give the graph of age-specific annual death rate (mathematically identical to $Z(t)$ and called by actuaries, 'the force of mortality') for the U.K. population in the 1960s. It is identical, in general form, to *Figure 3.1*.

Estimates of the parameters of the whole-life failure-probability profile of the constituent components (especially of the useful life, Phase II, mean failure rate λ) are an essential requirement for the prediction of system reliability. Additional information, such as repair-time distributions, then leads to estimates of availability, maintainability, and hence the level (and cost) of corrective and preventive maintenance.

3.3. Reliability Prediction for Complex Plants

For the purposes of analysis one way of regarding a large and complex industrial plant is as a hierarchy of parts ranked according to their function and replaceability (see *Figure 3.2*). At each functional level equipment may be connected either in series (as in *Figure 3.3*), in parallel (as in *Figure 3.4*), or in some combination of either. *Figure 3.5* is an illustrative example of such a model and its analysis.

The appropriate measure of reliability in the example of *Figure 3.5* has been taken to be the survival probability $P(t)$, as defined in Section 2.2.3. Its value at, say, 100 h has been calculated for the complete plant by analysing from the component level upwards, all

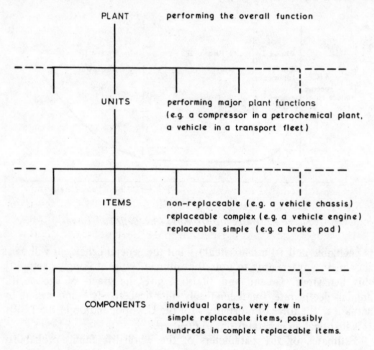

Figure 3.2. Plant hierarchy

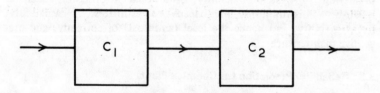

Figure 3.3. Series-connected components

the component mean failure rates being known or susceptible to estimation (for simplicity it is assumed that all components are in their useful-life phase, i.e. negative exponential p.d.f.s of times-to-failure). At each level the survival probability calculation takes the functional configuration into account. As already emphasised, such estimations

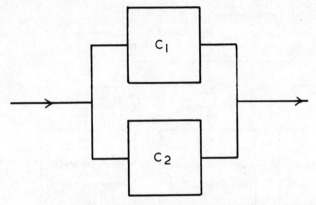

Figure 3.4. Parallel-connected components

of system $P(t)$ (in the example estimate $P(100 \text{ h}) = 0.50$) can be used in the selection of design or re-design alternatives, in the calculation of plant availability, or in the prediction of maintenance work load.

The simple, series and parallel, survival probability calculations used in *Figure 3.5* will now be outlined.

3.3.1 Series-connected Components (as in Figure 3.3)

For such a system (e.g. fuel pump feeding a carburettor) to work both components must work. Assuming that the failure behaviour of one component is quite uninfluenced by that of the other, i.e. that the failure probabilities are statistically independent, the survival probability, $P(t)$ for the system at time t is given by the product of the separate survival probabilities, $P_1(t)$ and $P_2(t)$, of the components at the time t, i.e.

$$P(t) = P_1(t) . P_2(t)$$

and for a system of n series-connected components

$$P(t) = P_1(t) . P_2(t) . P_3(t) . \ldots \ldots P_n(t)$$

Since survival probabilities must always be less than 100% it follows that *P(t)* for the system must be less than that of any individual component, a deduction which merely confirms what every engineer knows in his bones.

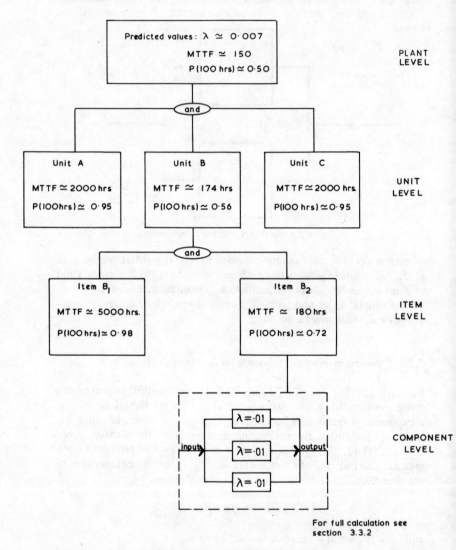

Figure 3.5. A plant reliability analysis

If both types of component behave according to the exponential p.d.f. of times-to-failure then

$$P(t) = P_1(t) \cdot P_2(t) = \exp(-\lambda_1 t) \exp(-\lambda_2 t) = \exp(-(\lambda_1 + \lambda_2)t)$$

and

$$f(t) = (\lambda_1 + \lambda_2) \exp(-(\lambda_1 + \lambda_2)t)$$

or in general

$$f(t) = (\lambda_1 + \lambda_2 + \ldots + \lambda_n) \exp(-(\lambda_1 + \lambda_2 + \ldots + \lambda_n)t)$$

So the overall p.d.f. of times-to-failure is also exponential.

Example. Four valves are connected in series, are statistically independent and each has a constant age-specific failure rate, λ of 0.0001 per hour.
   What is
   1. the survival probability at 1000 h running time?
   2. the average time to failure of the system?

1. For any one valve, $Z(t) = \lambda$, so $P(t) = \exp(-\lambda t)$ (see *Figure 2.11*).
For the system,

$$P(t) = (\exp(-\lambda t))^4 = \exp(-4\lambda t)$$

so

$$P(1000) = \exp(-4 \times 0.0001 \times 1000) = 67\%$$

2. For the system, $f(t) = 4\lambda \exp(-4\lambda t)$.

So

$$\text{Mean time-to-failure} = 1/(4\lambda) = 1/(4 \times 0.0001)$$

$$= 2500 \text{ h}$$

3.3.2 Parallel-connected Components (as in Figure 3.4)

Such a system fails only if both components fail. Assuming (as before) statistical independence, the probability $F(t)$ that both will *fail* before time t has elapsed is given by the product of the two separate failure probabilities, $F_1(t)$ and $F_2(t)$ i.e.

$$F(t) = F_1(t). F_2(t)$$

The system survival probability is therefore given by the expression

$$\begin{aligned} P(t) &= 1 - F_1(t). F_2(t) \\ &= 1 - (1 - P_1(t))(1 - P_2(t)) \\ &= P_1(t) + P_2(t) - P_1(t).P_2(t) \end{aligned}$$

Since survival probabilities cannot be greater than one hundred percent it follows from this last expression that the system survival probability must be greater than that of either of its parts. This again is a deduction that tallies with everyday experience; adding otherwise redundant parallel capacity is commonly done to improve reliability. Additional advantages commonly accrue in that, if the separate parallel units can be isolated, extensive preventive maintenance can be pursued with no loss in plant availability and, in the event of a failure, corrective maintenance can be arranged under less pressure from production or from competing maintenance tasks.

If units are already of very high reliability it follows that there are only very marginal increments in reliability to be gained by installing redundant capacity. By implication, such units are of high cost and the redundancy cannot therefore be justified other than in nuclear or highly toxic chemical plant, for example, where safety factors may be overriding. If high reliability units are used in stand-by parallel arrangement, (a) the reliability of the diverting valves, switchgear, and flow lines may become critical, and (b) steps have to be taken to ensure that the potentially little used units are functional. Texts such as Ref.3 treat the subject of high reliability in great detail.

Returning to the system (*Figure 3.4*) under analysis, if both types of component behave according to the exponential p.d.f. then

$$P(t) = \exp(-\lambda_1 t) + \exp(-\lambda_2(t)) - \exp(-(\lambda_1 + \lambda_2)t)$$

Now

$$f(t) = \frac{d}{dt}(P(t))$$

which in this case gives

$$f(t) = \lambda_1 \exp(-\lambda_1 t) + \lambda_2 \exp(-\lambda_2 t) - (\lambda_1 + \lambda_2) \exp(-(\lambda_1 + \lambda_2)t)$$

which is *not* a simple exponential p.d.f. The mean time to failure is given by the expression

$$\bar{t} = \int_0^\infty f(t).t.\mathrm{d}t$$

which in this case gives

$$\bar{t} = \frac{1}{\lambda_1} + \frac{1}{\lambda_2} - \frac{1}{\lambda_1 + \lambda_2}$$

Example. Three identical pumps are connected in parallel and are statistically independent. The system operates if at least one pump is working. The age-specific failure rate of this type of pump is of constant value, λ, of 0.01 per hour. Compare the survival probability of this system after 100 h with that of a single pump.

For the system, $\begin{aligned} P'(t) &= 1 - (1 - P(t))^3 \\ &= 1 - (1 - \exp(-\lambda t))^3 \\ &= 1 - (1 - \exp(-0.01 \times 100))^3 \\ &= 72\% \end{aligned}$

For a single pump

$$P(t) = \exp(-0.01 \times 100) = 37\%$$

3.3.3 Reliability and Preventive Maintenance

In examples such as that of *Figure 3.5* it was assumed that all components were used within their useful-life phase. Clearly, for reasons of cost or technology this will often not be the case in practice. Many components will have a useful life much less than the anticipated life of the system, e.g. the brake pads of an automobile. *Figure 3.6* is an elementary illustration of a situation of this kind, component A having a mean useful life less than that expected of the item, the reliability of which will be maintained only if A is replaced prior to failure.

Such replacements should
1. interfere as little as possible with the operation of other components,
2. not interrupt normal operation or production,
3. occur at intervals which exceed, as far as possible, the maximum operational cycle or production run.

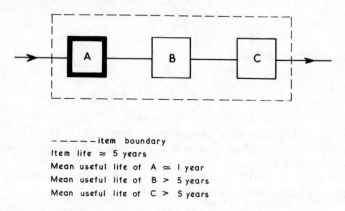

------ item boundary
Item life ≈ 5 years
Mean useful life of A ≈ 1 year
Mean useful life of B > 5 years
Mean useful life of C > 5 years

Figure 3.6. Preventive maintenance and reliability

The replacement of wear-out items and units will be discussed in Chapter 4.

3.4 Maintainability Prediction

As well as being able to design for a specified level of reliability it is also desirable to be able to estimate the time required, and resources needed, for maintenance. The cost of achieving a given level of availability can then be predicted. The basic method can be illustrated by reference once more to *Figure 3.5.*

Let the availability required of the plant be 0.95 and the predicted mean running time to failure (MTTF) of the system be 250 h.

If $\overline{M}$ is the mean repair time then

$$\text{Availability} = \frac{\text{MTTF}}{\text{MTTF} + \overline{M}}$$

i.e. in this case
$$0.95 = \frac{250}{250 + \overline{M}}$$

from which
$$\overline{M} = \frac{250}{0.95} (1 - 0.95) = 12.5 \text{ h}$$

Such a plant mean repair time, calculated to give a specified avail-ability, can be allocated at unit level and then sub-allocated at item level. These allocations will be meaningful only if repair time data is available (or can be realistically estimated) for the item level at least. The validity of the original calculation of $\overline{M}$ depends, of course, on the creditability of the estimate of system MTTF.

Analysis of this kind, used in conjunction with reliability data, enables alternative repair strategies to be compared in terms of cost and in terms of required resource spectrum. Unfortunately, there is a dearth of repair time data and such comparisons are therefore subject to great uncertainty. For a full treatment the reader should consult specialist texts such as those of Blanchard[4] and Green and Bourne[3].

3.5 Case Study 1: A Plant Availability Assessment*

Broadly speaking availability assessments are of two kinds:

1. Safety assessments concerned with protective systems — where failures are frequently unrevealed and show only when the system is tested or when demands are made on the system.
2. Plant availability assessments — where failures are revealed and result in some change in the plant operating condition. The failure is then repaired and the plant restored to its original operating state.

This case study illustrates the latter application.

3.5.1 The Concept of Availability

Availability is defined in a number of different ways by different industries; for example, in the chemical industry it is frequently defined in terms of lost production. In the simplest, two-state case we can assume that the plant is either working at full capacity or not working.

*Contributed by T.R. Moss of the National Centre of Systems Reliability.

Availability (A) can then be seen as the ratio of the operating time to the total time, i.e.

$$A = \frac{T_{up}}{T_{up} + T_{down}}$$

where the up-time is the time in the working state and the down-time the time in the non-working state. In this context, therefore, availability can be defined as a probability — the probability that the plant is working at any time t_i in the period considered; generally the interval between major overhauls — say one to two years.

If the plant is capable of working in more than two states — perhaps at an intermediate 50% state because of redundancy — the equation becomes slightly more complex. In these cases it is usual to define availability in terms of equivalent full-production time.

3.5.2 Methodology

A method used in NCSR for the assessment of chemical plant availability involves 10 steps, viz.

1. Definition of
   (a) assessment objectives
   (b) plant boundaries
   (c) inputs and outputs across boundaries
   (d) performance/failure criteria
   (e) availability model.
2. Appraisal of the plant's function and *modus operandi*.
3. Development of a block scheme, in hierarchical or flow sheet form, to a level at which reliability data are available.
4. Development of logic diagrams from which the overall availability model can be derived.
5. Rough optimisation of the data to be used in the model.
6. Carrying out the calculations.
7. Investigation of the accuracy of the reliability data for the high-scoring items and adjustment of the failure-rate and repair-time data if necessary.
8. Repetition of steps 6 and 7 until a constant solution is obtained.
9. Comparison of the results with other information (published reports, experience of plant engineers etc.).
10. Drawing conclusions and making recommendations.

3.5.3 Comments on Method

The steps which are particularly worthy of comment are 2, 3, 4 and 5 since they generally involve the largest proportion of project time.

Plant description In the majority of cases there is no complete description of the plant's function, or its modes of operation, available at the beginning of an assessment project. Separate descriptions of the plant sub-systems, particular equipments or operating procedures may exist in isolation but they are seldom put together into one comprehensive document. The same is generally true of the engineering drawings which are produced for engineers in particular disciplines, for example, as plant flow sheets, instrumentation diagrams etc.

The exercise of describing the plant is also a useful discipline in understanding the operation of the process. It is essential to ensure that the assessor and operator appreciate the basis on which the assessment is structured.

The block scheme The block scheme is generally a development of the heat and mass balance flow sheets. The detailed structure is determined by equipment redundancy considerations, sub-system interactions, storage availability, etc. The criterion is always to establish the lowest level of complexity consistent with the reliability data available. If a plant has one and only one configuration and we have reliability data available for this then the diagram consists of one block. This is seldom the case because of multiplicity of feed-streams, the combination of equipments in series, parallel or standby arrangements and other factors such as batch delivery of feedstock.

However, a real attempt must be made to maintain the simplest block structure consistent with a true representation of the plant.

Logic diagram From the block scheme failure logic diagrams are constructed to describe the logical relationships of all the various events which can affect plant availability. The analytic solutions obtained from these logic diagrams predict the availability of sub-systems from information of the reliability of the plant items, redundancy and storage considerations. Each plant item is characterised by a failure rate and a repair time; the product of these parameters giving a first order approximation of the item unavailability (downtime). For calculational convenience it is assumed that each item is in its useful-life, negative exponential p.d.f. phase.

The reliability data Particular care needs to be taken in selecting and estimating the reliability data for use in the availability equations. Tables[5, 6, 7, 8] of mean failure rates and repair times have been published. The National Centre also maintain a large data base with a useful subset in the chemical plant field. However, mean failure rates on their own are not sufficient; the effect of operating and environmental conditions needs to be estimated, also the proportion of the total failure rate that will cause a plant breakdown. It is in this area that considerable engineering judgement needs to be applied so that a coherent data set, truly representative of the plant being assessed, is assembled prior to the start of the calculations.

3.5.4 The Case Study

The assessment described here was concerned with a proposed large plant for the production of ammonia. Clearly the analysis cannot be discussed in detail but the procedure should become evident from the following brief summary. The procedure was outlined in Section 3.5.2 as a series of discrete steps but in practice there is, of course, considerable interaction, particularly between the first four stages.

Definitions The assessment objectives were defined in the original letter of agreement between NCSR and the client and included:

1. Where adequate repair and replacement policy information existed, estimates of the availability to be expected from the sub-system, main plant units and the plant as a whole.

2. Identification of areas of the plant and operating methods which would be critical to the availability of the plant.

3. Identification of areas where further and more detailed availability assessment could be required.

The plant boundaries were broadly defined by the battery limits although specific assumptions on the availability of feedstock, electric supplies, etc., were agreed with the client's engineers and clearly stated in the original assessment specification. This also effectively defined the inputs and outputs across the main plant boundaries.

The performance/failure criteria were defined, in a way specific to the client's requirements, as

1. failure to maintain 100% production (reduced output),
2. failure to maintain 50% production (shut-down).
(It should be noted that these are not mutually exclusive states.)

Adopting these failure criteria the overall availability model was developed around the concept of fractional dead time (which is a measure of the plant downtime). Fractional dead time (D) and availability (A) are complementary terms i.e.

$$A + D = 1$$

The fractional dead time indicates the proportion of time the plant is expected to be in the dead (i.e. non-working) state and is calculated from the expression*

$$D = \lambda \overline{M} / (1 + \lambda \overline{M})$$

where λ = unit failure rate
$\overline{M}$ = mean repair time.

The state fractional dead times are derived from the reduced output (RO) and shutdown (SD) logic diagrams and combined to give the overall fractional deadtime

$$D = \frac{D_{RO} + D_{SD}}{2}$$

A rigorous treatment of the mathematics is seldom warranted at this 'broad survey' level and it was therefore assumed that the plant fractional deadlines were the sums of the sub-system fractional dead times in each of the defined operating states. This was not strictly correct but the errors involved were considered to be well within the accuracy of the data used.

* Assume (i) equipment either running (up) or in repair (down), so total time, $T = T_{up} + T_{down}$;

      (ii) N failures, each of mean repair time $\overline{M}$, occur in time T so $T_{down} = N\overline{M}$,

and   (iii) λ is defined as the mean number of failures per unit of *running time* so

$$\lambda = \frac{N}{T_{up}} \; ; \text{or } T_{up} = \frac{N}{\lambda}$$

It follows that the fractional dead time is

$$D = \frac{T_{down}}{T_{up} + T_{down}} = \frac{N\overline{M}}{\dfrac{N}{\lambda} + N\overline{M}} = \frac{\lambda\overline{M}}{1 + \lambda\overline{M}}$$

Plant description The plant was designed to produce 1000 tonnes per day of ammonia using coal as feedstock. By-product outputs would be hydrogen sulphide and carbon dioxide. Liquid oxygen would also be produced as a result of surplus capacity of the air-separation plant. The plant would produce its own steam for the process and the turbines, oxygen and nitrogen for the process, and air for instruments. Waste products would be ash, small quantities of nitrogen and carbon dioxide, and other impurities. Some of the gaseous waste would be vented to atmosphere and some returned to the boilers for burning.

Inputs to the plant would be

(a) raw coal for the process and steam raising,
(b) electric power, for auxiliary and standby pumps, conveyors and a small number of heaters,
(c) drinking water, for boiler feed make-up,
(d) purified sewage, for cooling water and wash-water make-up,
(e) methanol, for make-up in the gas-purification system,
(f) fuel-oil, for gasifier start-up,
(g) sulphuric acid and caustic soda, for regeneration of ion and cation exchangers in the feed-water treatment plant,
(h) hydrazine and sodium phosphate, for boiler feed-water treatment.

Economy of heat would be practised throughout the plant. Regenerative heat exchangers using the heat generated during a process would be extensively used to heat incoming and cool outgoing gas. In many cases series heat exchangers, using heat sources from different stages in the unit process, would be used to maintain the correct heat balance. This would increase the efficiency but would obviously also increase the complexity of the plant.

The plant was essentially a single-stream system involving 22 units some of which would have redundancy. A simplified block scheme is shown in *Figure 3.7*. Raw coal is fed to the steam raising plant and the coal preparation plant. In the latter the coal is pulverised and the coal dust then passed pneumatically into the gasifiers, where it is mixed with oxygen (from the air-separation plant) and steam. The mixture is burned, producing raw synthesis gas comprising CO, H_2 and CO_2 with small quantities of impurities. The dust is removed from the gas in disintegrators and electrostatic precipitators before the first stage of compression to 55 bar in two steam-driven, centrifugal, raw-syngas compressors operated in parallel.

After compression the raw syngas passes through a complex clean-up process to remove impurities and CO_2 and then pure nitrogen is added to attain the correct stoichiometric composition. The second stage of

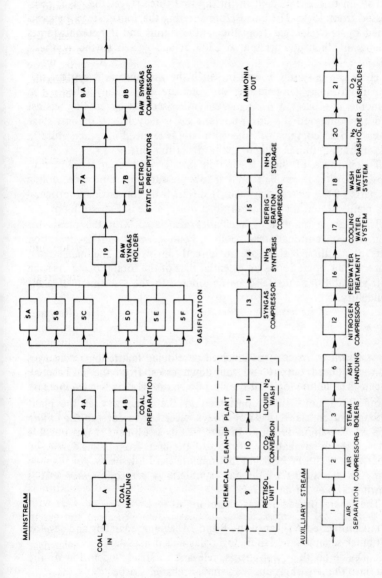

Figure 3.7. Ammonia plant, simplified block scheme

compression involves a multi-stage centrifugal compressor driven by two steam turbines in tandem. In the first three stages the gas is compressed from 45 to 212 bar. Before entering the fourth stage the gas is joined by re-cycled gas from the synthesis unit and the combined gas streams are then compressed to 226 bar and passed to the synthesis unit.

There is one synthesis unit in which the gas mixture is synthesised, in part, at high temperature and pressure over catalyst beds. The ammonia is separated from the unsynthesised gas by cooling against cold ammonia and the unsynthesised gas is recycled to the last stage of the syngas compressor. The ammonia is reduced in temperature in a flash drum before passing to the refrigeration unit.

The purpose of the refrigeration unit is to cool the product to − 37°C so that it can be stored at low pressure and to provide cooling for parts of the clean-up system. It differs from conventional refrigeration plants in that the refrigerant does not operate in a closed loop.

There are a number of ancillaries associated with the plant, for example, conventional steam boilers, cooling water and air-separation units, etc. Although these are not directly involved in the main stream process their operation is an essential part of the total process. As such their availabilities must also be considered in the assessment of plant availability.

The assessment Assessment involved developing failure logic diagrams, for the reduced output and shut-down cases, from the main block scheme. The failure logic for the shut-down case is shown in *Figure 3.8*.

Individual units were considered in the same way as the plant although, of course, it was not always necessary to develop the failure logic in diagrammatic form. In fact, in this assessment it was possible to calculate the availability of most of the units from tables developed directly from the Process and Instrumentation Flowsheets. An example from the assessment of the raw synthesis − gas compressor unit is shown in *Tables 3.1* and *3.2*.

The general procedure was, therefore, to

1. Consider each sheet of the P and I Diagram separately and develop unit block schemes and failure logic diagrams if necessary.
2. Divide all the components, relevant to steady operation of the unit, into the largest groups with similar reliability aspects.
3. Tabulate the estimated reliability parameters and calculate the dead-times for each component of the group and sum to determine the

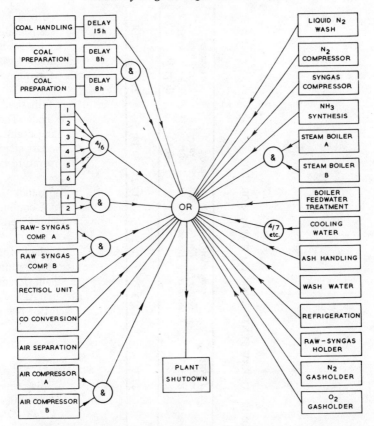

Figure 3.8. Failure logic diagram – shut-down. (Failure to maintain 50% production)

average failure rate and repair time. (The average repair time is assumed to be the average dead-time per fault).

4. Tabulate the calculated failure rates and dead-times of each group, together with those of the components which did not merit special consideration, and sum to determine the total failure rate of all components on that particular sheet of the P and I Diagram.

5. Sum the failure rates and dead-times for each P and I Diagram to give the totals for the whole unit.

In some cases it was not necessary to sub-divide into groups, for example, when there was no redundancy of components. In such cases steps 2 and 3 above could be omitted.

Table 3.1. COMPLETE RAW SYNTHESIS-GAS COMPRESSOR UNIT

Equipment	No. off	Failure rate, f/y	Average repair time, h	Redundancy Reduced output	Redundancy Shut-down	Average failure rate RO (f/y)	Average failure rate SD (f/y)	Average dead-time RO (h/y)	Average dead-time SD (h/y)
Raw-Syngas Compressor Set	2	8.64	17.1	2/2	1/2	17.28	0.14	295.5	2.41

Table 3.2. SINGLE COMPRESSOR SET . Drg. M.60.90301260b

Equipment	No. off	Failure rate, f/y	Average repair time, h	Redundancy		Average failure rate		Average dead-time	
				Reduced output	Shut-down	RO (f/y)	SD (f/y)	RO (h/y)	SD (h/y)
Steam turbine	1	0.6	70	1/1	1/1	0.6	0.6	42	42
Compressor casing	3	0.54	30.5	3/3	3/3	1.62	1.62	49.41	49.41
Lubrication system	1	1.07	4.79	1/1	1/1	1.07	1.07	5.14	5.14
Gas coolers	5	0.01	72	5/5	5/5	0.05	0.05	3.6	3.6
Ejector	2	0.1	24	1/2	1/2	negl.	negl.	negl.	negl.
Pneumatic valves	4	0.35	4	4/4	4/4	1.4	1.4	5.6	5.6
Pressure-relief valve	3	0.5	4	3/3	3/3	1.5	1.5	6.0	6.0
Hand valves	40	0.02	4	40/40	40/40	0.8	0.8	3.2	3.2
Non-return valves	2	0.5	4	2/2	2/2	1.0	1.0	4.0	4.0
Gear seal	1	0.6	48	1/1	1/1	0.6	0.6	28.8	28.8
			17.1		Total	8.64	8.64	147.75	147.75

Note: The figures in the Redundancy columns show the ratio of equipments required to equipments available to maintain operation above the state-heading.

Estimating failure rates Mention has already been made of the care required in selecting and estimating the reliability parameters. Of these the failure rate is most prone to variation. It is important that the consequences of failures are fully explored in conjunction with information on the failure modes associated with particular faults. On mechanical valves, for example, the major failure mode is external leakage; the consequences of such a leak are obviously different for flammable synthesis gas and water. The estimates of the appropriate failure rate, may, therefore, vary quite considerably from system to system.

Table 3.3.

Equipment	Average failure rate, f/y	Average repair time, h
Heaters, electric	0.02	72
Ion columns	0.1	24
Motors, electric		
Small	0.03	4
Large	0.12	148
Pressure vessels	0.001	72
Pumps		
Centrifugal and general	2.6	24
Oil	0.5	8
Scrapers	0.7	8
Stirrers	0.03	4
Tanks	0.001	72
Transformers, high voltage	0.004	24
Turbines		
Nitrogen	0.4	70
Steam	0.6	70
Valves		
Hand-operated	0.02	4
Motorised	0.05	4
Non-return	0.5	4
Pneumatic	0.35	4
Pressure relief	0.5	4
Solenoid	0.08	4

At the same time it may be worth sacrificing some degree of accuracy in the interest of simplicity. In this case it was estimated that the variation in operating conditions was unlikely to give data variations much greater than a factor of two or three and the benefits of using one data set for application to all units outweighed the other factors, at least for the first estimates.

An extract from the table of reliability parameters used in the assessment is shown in *Table 3.3*.

The plant availability calculation Since aspects of redundancy and storage capacity were taken into account in the individual unit assessments the final calculations involved a simple summation of the unit downtimes for each defined failure state. Clearly, however, expressions can be developed directly from the plant failure logic diagrams, if preferred.

To illustrate the point consider the calculations for Unit 8 – the Raw Syngas Compressors on which data are presented in *Tables 3.1* and *3.2*. Sections of the plant logic diagrams concerned with the Raw Syngas Compressors, for the reduced output and shutdown states are shown in *Figure 3.9*. Either compressor failing produces a reduced output state, the fractional dead time for Unit 8 is then

$$D_{RO8} = D_A + D_B$$

where
$$D_A = D_B = \frac{\lambda \overline{M}}{1 + \lambda \overline{M}}$$

$$\lambda = 8.64 \text{ f/y}$$
$$\overline{M} = 17.1 \text{ h}$$
$$\text{so } \lambda \overline{M} = 0.017$$
$$\text{and } D_{RO8} = 2 \times 0.017$$

The reduced output deadtime is therefore

$$2 \times 0.017 \times 8760 = 295 \text{ h per year (cf. } \textit{Table 3.1}\text{)}.$$

Both compressors need to fail to produce a shutdown state and the fractional deadtime for Unit 8 is then:

$$D_{SD8} = D_A \times D_B$$

$$= (0.17)^2$$

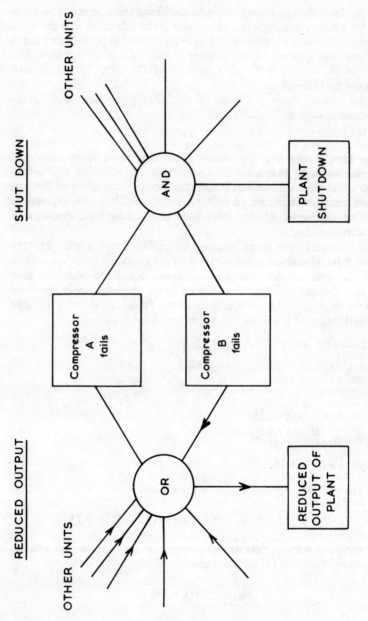

Figure 3.9. Section of plant logic diagram, Unit 8, Raw Syngas Compressors

The shutdown deadtime in this case is therefore

$$(0.017)^2 \times 8760 = 2.4 \text{ h per year}$$

3.5.5 Conclusions

For commercial security reasons it is not possible to give further details of the actual assessment carried out, except in general terms. The assessment indicated that rotating machines were likely to be responsible for over 40% of the downtime with three other main equipment types accounting for a further 40%. Certain sections of the plant were identified — the instrumentation and the gasification and chemical clean-up units — as particular areas where further work was required.

The main benefit of the assessment was as a comparative study to identify the sensitive areas of the plant. However, it is worth noting that the final estimate of downtime closely corresponded to figures quoted for ammonia plants operating in the U.S.A.

REFERENCES

1. Lomnicki, Z. A., 'Statistical Approach to Reliability', *J. Royal Statistical Society,* A, 1, 136 (1973)
2. Bazovsky, Igor, *Reliability Theory and Practice,* Prentice Hall (1961)
3. Greene, E. and Bourne, J., *Reliability Technology,* Wiley (1972)
4. Blanchard, S. B., *Logistics Engineering and Management,* Prentice Hall (1974)
5. Johnstone, D. E. and Merver, D. T., *Summary of Component Failure Rate and Weighting Function Data, etc.,* WADC Tech. Report 57–668. Astia DDC, AD–142120
6. MIL Handbook 217A, *Reliability Stress and Failure Rate Data for Electronic Equipment*
7. Dumner, G. W. A., 'Reliability Chart', *Electronic Components,* Sept. (1967)
8. Anyakora, S. N., Engel, G. F. M., and Lees, F. P., 'Some Data on the Reliability of Instruments in the Chemical Plant Environment', *The Chemical Engineer,* Nov. (1971)

Chapter 4

Maintenance Planning

4.1 Introduction

It was shown in Chapter 3 that complex plant can be divided, according to function and replaceability, into three distinct levels. The delegation of responsibility for the replacement and repair decisions of a given level differs between plants but higher management usually has responsibility for replacement of *units* (or, indeed of the manufacturing plant itself) and maintenance management for replacement and repair of *items* and below (see *Figure 4.1*). This division of responsibility is obligatory

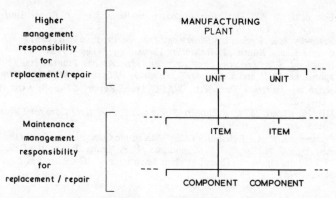

Figure 4.1. The relationship between plant structure and maintenance decision making

because replacement strategy for units is influenced by external factors (mostly long term) such as obsolescence, sales, or cost of capital, as well as internal factors (mostly short term) such as maintenance cost and operating cost. Consequently the replacement of units (*replacement*

70

strategy) can be considered as a part of the corporate strategy. However a shorter term *maintenance plan* is needed for the maintenance of units by the adoption of appropriate *maintenance policies* (e.g. repair, replacement, modification, etc.) for the constituent items and components. Strategy and plan are inter-related because maintenance cost influences unit replacement which in turn affects the maintenance plan. This chapter will mainly be concerned with the maintenance plan but will conclude, however, with a brief discussion of replacement strategy.

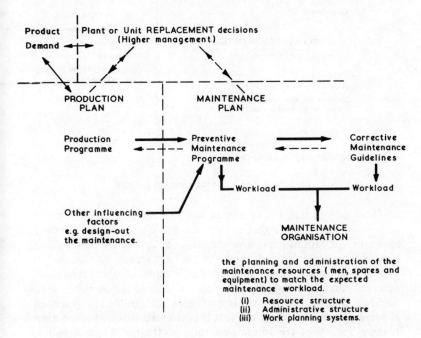

Figure 4.2. A systematic approach to formulating a maintenance plan

The *maintenance plan* should lay down a rational basis for formulating a programme of preventive maintenance and should provide guidelines for corrective maintenance. The *maintenance organisation* needed to execute the maintenance work load is dependent upon the maintenance plan. A procedure for the formulation of the plan and of the organisation is shown in *Figure 4.2*. For clarity, the functions referred to in the figure will be examined separately (preventive and corrective maintenance in this chapter and organisation in Chapter 5) but in practice they must be regarded as inter-related activities.

4.2 Maintenance Policies, Preventive and Corrective

There are a number of maintenance policies that can be specified, individually or in combination, for each unit of plant (see *Figure 4.3*). The rationalised sum of such specified policies for the whole manu-facturing plant constitutes the maintenance plan.

```
(I)    FIXED  TIME  MAINTENANCE
       Individual  or  group
       replacement etc.                         ⎫
                                                ⎪  For complex items
                                                ⎪  and / or continuously
(II)   CONDITION - BASED - MAINTENANCE          ⎬  operating plant
       Continuous  or  periodic  etc.           ⎪  consider
                                                ⎪  (IV) CPPORTUNITY
                                                ⎪       MAINTENANCE
(III)  OPERATE - TO - FAILURE                   ⎪
       Corrective maintenance by repair         ⎪
       in situ or by replacement  etc.          ⎭


(V)    DESIGN - OUT - MAINTENANCE
```

Figure 4.3. The major maintenance policies

Those actions that can be carried out before failure can be regarded as preventive and those that are carried out afterwards as corrective. Since, by definition, preventive maintenance actions are deterministic they can be scheduled and usually carried out separately according to a *preventive maintenance programme*. Because of the probabilistic nature of failure, and the uncertainty surrounding corrective maintenance decision making, corrective maintenance cannot be programmed. However, for critical units of plant it is essential that *corrective main-tenance guidelines* are formulated for maintenance decision making after failure.

Before considering a procedure for determining the 'best' maintenance plan the policies listed in *Figure 4.3* must be looked at in some detail.

4.2.1   Fixed-time Replacement or Repair, prior to Failure

This is only effective where the failure mechanism of the item is clearly time dependent, the item being expected to wear out within the life of the unit, and where the total costs (direct and indirect) of such replace-ment are substantially less than those of failure replacement-repair, i.e. the item could be classified as simple-replaceable.

The principle of fixed-time replacement is illustrated in *Figure 4.4*[1]. *Figure 4.4(a)* shows an item failure distribution which is clearly time dependent. The problem is to find that fixed-time replacement period which minimises the sum of fixed-time and failure replacement costs.

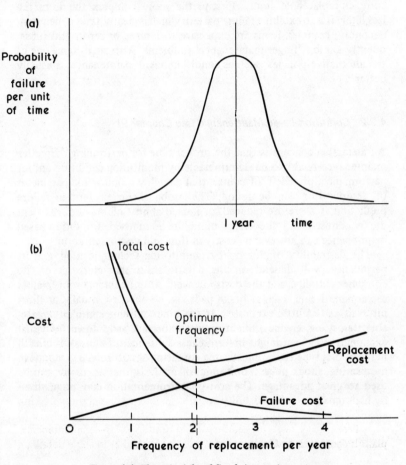

Figure 4.4. The principle of fixed-time maintenance

The data of *Figure 4.4(a)* is used in a calculation of the probable failure rate as a function of replacement period. This is then combined with cost data to give total expected cost rate as a function of replacement period (*Figure 4.4(b)*). The indicated optimum-frequency periodic replacement will only be the 'best' maintenance policy if it is of lower

cost than any of those alternative policies listed in *Figure 4.3*. The cost
and difficulty of collecting the requisite statistical data (see Chapter 2)
is also a factor to be taken seriously into account.

There are two main reasons why such a policy is inappropriate for
complex-replaceable items. Firstly, the more complex the item the
less likely it is to exhibit a failure pattern which is clearly time dependent.
Secondly, complex items are expensive to replace or repair and subsequently exhibit 'finger maintenance' problems. With such items one of
the alternative policies, such as condition-based maintenance, is usually
better.

4.2.2 Condition-based Maintenance (see Chapter 9)

An attractive concept is that the proper time for performing corrective
maintenance ought to be determinable by monitoring condition and/or
performance, provided, of course, that a readily monitorable parameter
of deterioration can be found. The probabalistic element in failure
prediction is therefore reduced, or indeed almost eliminated, the item
life maximised and the effect of failure minimised. Condition-based
maintenance can however be costly in time and instrumentation.

The desirability of this policy, monitoring technique used, and its
periodicity, will depend on the deterioration characteristics of the
equipment studied and the costs involved. At one extreme some simple
replaceable items, such as brake pads can be checked visually at short
intervals and at little expense*. At the other extreme complex replaceable items, e.g. engines, might require expensive strip down for visual
examination (which might in itself cause subsequent failures). It is with
items of this type that sophisticated condition monitoring, e.g. vibration
monitoring, shock pulse monitoring, oil analysis, thermography, can be
used to great advantage. The cost of instrumentation may be justified
by high repair and unavailability costs.

4.2.3 Opportunity Maintenance

This term is used for maintenance actions, taken after failure or during
fixed-time or condition-based repair, but directed at items other than
those that are the primary cause of the repair. The policy is most
appropriate for complex replaceable or continuously operating items of

---

* That is, traditional inspection routines which can be carried out separately, or in
conjunction with periodic servicing.

high shut-down or unavailability costs and, typically, might take the form of operation to failure and specification of critical items to be dealt with at that time.

4.2.4 Operation to Failure and Corrective Maintenance

No predetermined action is taken to prevent failure. The emphasis might well be on efficient corrective maintenance.

Corrective maintenance arises not only when an item fails but also when indicated by condition-based criteria. The basic task is establishment of the most economic way of restoring the unit to an acceptable condition. For example, for a failed complex replaceable item the alternatives might be as follows.

Repair in situ. Stripping down in the operating location and replacing the defective components. This might incur plant or unit unavailability.
Replacement of the whole item by a new or reconditioned item. This minimises unavailability. The removed item can be repaired, reconditioned, or scrapped at the maintenance base.

Many factors influence the repair-replace choice, the most important being the cost of unavailability, the time of repair compared with that of replacement, the availability and cost of resources. Such factors are continually changing and this, together with the many possible causes for the defect and the many possible methods of repair, means that a corrective maintenance plan can only provide a framework to assist decision making[2]. This dynamic nature of the problem is by far the most important feature of maintenance decision making which, therefore, cannot be tackled without an information system, as outlined in *Figure 1.4*, available to the right people at the right time. The information thus provided can only be used profitably if the decision makers, at their respective levels, have a thorough understanding of the plant for which they are responsible.

4.2.5 Design-out Maintenance

By contrast with the preceding policies, which aim to minimise the effect of failure, design-out maintenance aims to eliminate the cause of maintenance. Clearly, this requires engineering action rather than maintenance action but it is often part of the maintenance department's responsibility.

This is usually a policy for areas of high maintenance cost which exist either because of poor design or because the equipment is being used outside its design specification. A plant-condition control system of the type outlined in *Figure 1.4* will enable such areas to be identified and the choice is then between the cost of re-design or the cost of recurring maintenance.

4.3   The Determination of a Maintenance Plan

The maintenance plan for a plant should be built up by selecting for each unit, the best combination of the policies outlined in *Figure 4.3* and then by co-ordinating these policies in order to make the best use of resources and time.

Ideally, the preventive and corrective actions for each unit of plant should be specified in some detail by the manufacturers. This usually is the case for the simple replacement items where maintenance is inexpensive and deterministic but is rarely so for complex replaceable items where maintenance is costly and probabilistic. Items in the third category, non-replaceable, should, by definition, need no pre-determined maintenance action since their expected life should exceed that of the plant. Certain critical items in this category might, however, benefit from periodic condition-based maintenance.

Many factors affect selection of the policy appropriate for each item and this, together with the large number of items usually involved, gives rise to the need for some systematic procedure for determining the best maintenance programme for a particular period of time. The stages of such a procedure will now be outlined.

4.3.1   Classification and Identification of Equipment

This is important but usually tedious and difficult because of complexity and size. It is suggested that the classification into units and items should be based on replaceability and function, as outlined in *Figure 3.2*. The simplest of the various systems for identification of plant is probably some form of numerical coding.

4.3.2   Collection of Information

Acquisition of all information which might be relevant to maintenance planning is essential for every unit of plant. Since maintenance is inseparable from production it is inevitable that the information of greatest relevance is the production pattern (e.g. is it continuous, fluctuating or intermittent?) and the nature of the process (e.g. how much plant is redundant?). Such information having been obtained, it is then possible to construct a schedule, for each unit and each decision period, of the expected time available for maintenance which will not involve production loss.

Other information (much of which may be provided by the manufacturer) that *may* be required for each item is as follows:

Manufacturer's maintenance recommendations	:	Actions, periodicities, etc.
Equipment factors (which assist prediction of maintenance work)	:	Failure characteristics: mean-time to failure; minimum life; failure mode. Repair characteristics: mean-time for repair; time after failure before plant function affected; level of redundancy.
Economic factors (which assist prediction of main-critical units)	:	Consequences of failure; cost of replacement prior to failure, item material cost; monitoring cost.
Safety factors (which place constraints on the decision)	:	Internal; environmental; statutory regulations[5].

4.3.3 Selection of Policy

The 'best' policy for each item can be determined by first identifying the policies which are effective and then deciding which is the most desirable. The choice will depend on many factors and the decision criterion is normally one of minimum cost given that safety criteria, statutory and otherwise, can be met. It is advantageous to apply this selection process separately to the different categories of item.

Simple-replaceable. A detailed schedule of actions, periodicities and resources required recommended by the manufacturer. Most often, the problem is how to best programme the large number of different actions (for the whole plant) in order to co-ordinate resources and then match them to scheduled down-time.

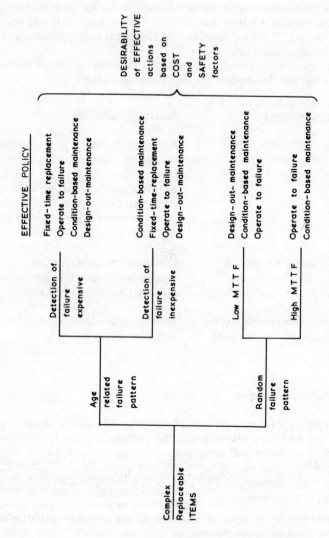

Figure 4.5. Selection of maintenance policy for complex plant items

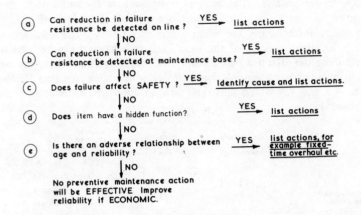

Figure 4.6. Decision chart – assessment of the potential effectiveness of maintenance actions

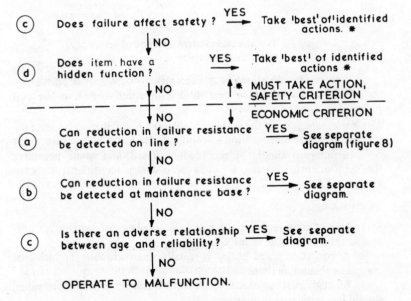

Figure 4.7. Decision chart – assessment of the desirability of identified effective maintenance policies

Complex-replaceable. The principal equipment, safety and cost factors can be ranked in order of importance and usually this is all that is necessary for selecting the best maintenance policy. *Figure 4.5* shows that the characteristics of the item's failure behaviour can be used to determine the effective policies and that the safety and cost factors can then be used to determine the most desirable policy (see also *Figures 4.6, 4.7* and *4.8* which illustrate such an approach as used by some airlines[3]).

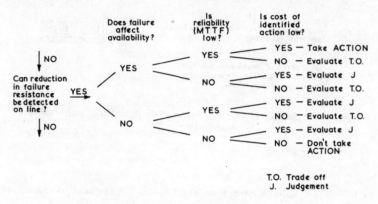

T.O. Trade off
J.   Judgement

Figure 4.8. Decision chart — determination of best policy

In only a few situations is it necessary to construct sophisticated statistical, mathematical, or cost models in order to deduce the best policy[4].

Non-replaceable. Since these are not expected to fail it must be assumed that no predetermined action is required. However in the rare event of such failure it should be recorded, analysed, and where necessary the appropriate maintenance policy or re-design identified (as shown in *Figure 4.5*).

Summarising:

1. A fixed-time replacement policy is usually most suitable for low-cost simple-replaceable items.

2. A condition-based policy is usually most suitable for high-cost complex-replaceable items.

3. All high cost maintenance items, replaceable or non-replaceable, should be considered for designing out.

4. Where no preventive maintenance/design-out action is effective, or desirable, the item is operated to failure.

4.3.4 Preventive Maintenance Programme

When the individual analyses are complete the aggregate actions and periodicities must be examined for opportunities of co-ordination (by joint scheduling, into set periods, of all actions on a group of items, or on a unit). This will involve a compromise between the individual optimum schedules, the most economic use of labour, and maximum plant availability. These set periods must carry a time tolerance to allow for contingencies such as uncertainty of production schedule. Inspection routines, lubrication routines, servicing schedules, overhaul schedules result from this analysis.

4.3.5 Corrective Maintenance Guidelines

When the plant is new it is difficult, even after the above type of analysis, to forecast the level and nature of the corrective maintenance load. During the early life of the plant the forecast is uncertain and will rely heavily on manufacturers' information and the plant engineers' experience. Obviously this forecast will improve during the life of the plant and the corrective maintenance load can be planned for with more certainty. The critical decision in this respect is the level of spares that is to be carried. The greater the number of such plant items carried the lower is the unavailability cost on failure, and the easier it is to schedule corrective maintenance; on the other hand the 'holding costs' are higher. The maintenance manager has the problem of minimising the sum of these costs which means it is essential to identify the critical units/items of equipment and to ensure that the best corrective maintenance plan is adopted.

4.3.6 Maintenance Organisation

The maintenance organisation needed to carry out this expected maintenance work load involves three main areas of decision making: the resource structure, the administrative structure, the work planning systems. This difficult management problem is covered in detail in the next chapter.

4.3.7 Illustrative Examples

These are drawn from the case study of Chapter 13 and are concerned with the diesel-powered loaders discussed therein.

Example A concerns the diesel engine air filters which were simple replaceable items. The filters contained oil, which had to be kept within prescribed limits for correct filter functioning and could be regarded as a separate item. Failure of the filter in the dusty atmosphere of the mine caused failure and scrapping of the engine. The predominant cost factors were

> the high cost of filter failure,
> the low cost of oil and filter replacement.

The selected preventive maintenance policy was (a) operator oil check at the beginning of each shift, and (b) fixed time-limit filter replacement well within the filter life.

This policy was not successful because action (a) was not carried out sufficiently reliably and the consequences were severe. The filter was therefore redesigned as a dual unit so that if the main filter failed a secondary filter would provide short term (longer than one shift) protection and give an indication of the primary filter failure. The modified preventive maintenance policy was

> (a) operator oil check at the beginning of each shift,
> (b) operator monitoring for primary filter failure,
> (c) maintenance check for primary filter failure after each shift, and replacement as necessary.

Example B concerns the loader engines which were complex replaceable items. (This did not include items which were engine parts but could be replaced separately e.g. fuel injectors.) The predominant equipment factor in this case was the relatively short periodic service-time which was needed for the fixed-time replacement of fuel injectors and other associated items. This, in conjunction with the long term wear-out characteristics of the main engine parts, and the high cost of engine replacement led to the adoption of a preventive maintenance policy of carrying out condition checks for engine wear at monthly service periods (or at some recorded service time), once past the first three months of operation.

Corrective maintenance action was necessary as a result of these checks and as a result of engine failure during operation. A repair-versus-replacement decision tree was constructed for the main faults and an information system set up to provide the decision maker with the necessary up to date information. In addition the level of components, engines and labour, appropriate for the maintenance work, had to be organised. This example illustrates the close inter-dependence of preventive maintenance, corrective maintenance, and organisational decision making.

4.4 Replacement Strategy

In Section 4.1 reference was made to the possibility of estimating a minimum cost replacement frequency (fixed-time replacement prior to failure) for units of plant using replacement theory. Calculating the optimum time to replace a privately owned motor car is probably the best known problem of this kind. Such replacement problems have provided one of the happy hunting grounds of the operational research theorists; the very elaborate mathematics of highly hypothetical situations has been developed, in some cases, for the intellectual pleasure therein and not for its utility. The same thing has happened with queueing and inventory problems (discussed in Chapters 6 and 7).

In the real world replacement decisions have to take account of a multitude of considerations: running, repair and acquisition costs (and their variation), likely income from sale, failure pattern, opportunities provided by seasonal or statutory shut-down, re-conditioning possibilities, fiscal considerations (tax incentives, import duty, etc.), cost of borrowing, obsolescence, alternative investment etc. The greater the number of such considerations that a replacement calculation includes, the very much greater is the complexity of the algebra. Replacement models only take account, therefore, of a few of the more important variables in any particular case and are, to that extent, always an approximation. Also, it is the authors' opinion that such models are of very limited applicability. Industrial replacement problems are usually dominated by one or two quite clear restrictions which point unambiguously to the policy required; in addition, changes in external factors can be large, sudden and unpredictable.

Nevertheless, the following highly simplified example is offered as an illustration. It is of a deterministic nature in which averaged costs and trends are fairly predictable, as might be the case with a substantial unit of capital equipment.

Example. A fixed-time-replacement model for a unit of plant.

When new, the equipment's operating cost is 0 (£/year). This rises linearly with time at a rate i (£/year/year) so that after n years the operating cost would be $0 + ni$ (£/year) and averaged over that time the mean annual operating cost would be $0 + (ni/2)$ (£/year).

If the equipment has cost £A to begin with, and were to be sold after n years for £S, the mean annual cost in this respect would be $(A - S)/n$ (£/year).

There will also be the cost of raising the above money which could have been done by borrowing £$(A - S)$ for n years (repaying this in annual instalments over the period) and borrowing £S for n years (repaid at the time of sale). Assuming simple interest at rate r, and

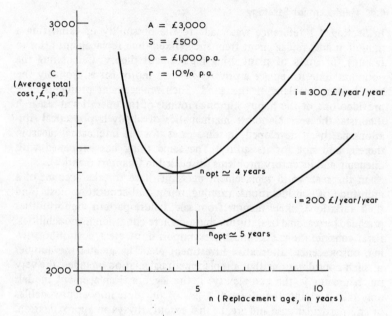

Figure 4.9. Plot of average cost, vs. replacement age, model in Section 4.4

remembering that the amount borrowed decreases steadily, the mean cost of the first item would be $(A - S)r/2$ (£/year); the mean cost of the second item would be Sr (£/year). The total mean borrowing cost would therefore be the sum of these two, which is $(A + S)r/2$ (£/year).

So, the mean annual cost if the equipment were to be replaced after n years would be

$$C = 0 + \frac{ni}{2} + \frac{A - S}{n} + \frac{(A - S)r}{2}$$

The object is to find the value of n which minimises C. At this value

$$\frac{dC}{dn} = \frac{i}{2} - \frac{A - S}{n^2} = 0$$

i.e.

$$n = \left[\frac{2(A-S)}{i} \right]^{1/2}$$

If A = £3000, S = £500 and i = 200 (£/year/year) then the optimum replacement age is

$$n = \left[\frac{2(3000 - 500)}{200} \right]^{1/2} = 5 \text{ years}$$

Instead of the above mathematics a graphical method could have been used. The advantage of a graph is that the importance of the optimum, and its sensitivity to changes in assumed data, can be clearly appreciated. In *Figure 4.9*, C is plotted against n, revealing that C changes little between, say, n = 4 and n = 8 years, and also that if the assumed value of i is increased by 100 (£/year/year) the resulting optimum replacement age is lowered by about a year.

Models like this can be developed into as sophisticated a form as is desired; the next step, for instance, might be to include discounted cash flow costing. Changes in predicted costs and earnings patterns might be met by the adoption of dynamic programming models but the cost of this would have to be justified. Special methods have been developed for certain cases, e.g. the repair limit method[2] for vehicle fleets.

REFERENCES

1. Jardine, A. K. S., *Maintenance, Replacement and Reliability,* Pitman (1973)
2. Hastings, N. A. J. and Thomas, D. W., 'Overhaul Policies for Mechanical Equipment', *Proc. Instn. Mech. Engrs.,* **185,** 40/71 (1971)
3. Nowlan, F. S., 'Decision Diagram Approach to on-Condition Philosophies', *Aircraft Engineering,* Mar. and Apr. (1972)
4. Jardine, A. K. S. and Kirkham, A. C. J., 'Maintenance Policy for Sugar Refinery Centrifuges', *Proc. Instn. Mech. Engrs.,* 18753/73 (1973)
5. Mitchell, E., *The Employers' Guide to the Law on Health, Safety and Welfare at Work,* Business Books Ltd. (1975)

Chapter 5

Organisation of Maintenance Resources

5.1 Introduction

Organisation of maintenance resources for a fluctuating, multi-trade, maintenance work load is a difficult managerial problem which involves three areas of organisational decision making, i.e. those concerned with

1. The mix, location and size of the maintenance resources.
2. The nature and type of administrative procedure.
3. The type of work planning and scheduling system necessary to match the level and supply of resources to the work load.

The general principles of organisation theory will not be discussed in detail since they are dealt with in many other books. The discussion will therefore be confined to the particular problems of maintenance organisation.

Maintenance organisations can take an infinite number of forms, the best for a particular situation being determined by systematic consideration of the influencing factors, many of which are inter-related. Also, as already emphasised in Chapter 4, the maintenance problem is dynamic and the organisation will require modification as the influencing factors change.

5.2 Maintenance Resource Structure

The objective is to set up that mix, location and size of resources which will best respond to the expected maintenance workload*. In order to

---

* The key resource in this respect is the maintenance tradeforce, spares and equipment being secondary but closely related aspects. The organisation and control of spares will be discussed in Chapter 7.

accomplish this it is necessary to understand the general characteristics of the workload and to have adequate knowledge of the maintenance resources. The basic problem is balancing the cost of unavailability against the utilisation of maintenance resources; *the most important single influencing factor is the cost of unavailability.*

The workload falls into two categories:

1. *The deterministic load.* That which can be both planned and scheduled in the long term, e.g. preventive maintenance, modifications projects, and a proportion of the corrective maintenance.

2. *The probabilistic load.* That which can be scheduled only in the short term, e.g. corrective and emergency maintenance. At best only the average level is known.

The fluctuating nature of the latter, which also incurs a level of indirect cost, is the major organisational difficulty.

5.2.1 Tradeforce Mix

As well as being classified as above the workload must be divided according to trade, e.g. fitting, instrumentation, electrical, etc. Such constraint on labour flexibility, caused either by work complexity or by trade demarcation rigidity, greatly influences the range of plant equipment that can be effectively dealt with by an individual tradesman, increases the number and size of gangs, makes more difficult the achievement of high labour utilisation, and renders work planning more difficult owing to the complexity of trade co-ordination.

Flexibility, particularly with regard to simpler maintenance work, can be improved in two principal ways. Firstly through formal training programmes, both internal and external, to broaden tradesmen's skills and secondly through productivity agreements and other management-union bargains.

5.2.2 Tradeforce Location

The most important factors in this area are the plant layout and the level and cost of plant unavailability. If the work load is widely distributed, and contains a large proportion of high cost emergency maintenance, permanent decentralisation of the trade force reduces the costs, both direct and indirect (due to plant unavailability), of travelling. Conversely, for a localised work load centralisation of the maintenance tradeforce improves utilisation of resources. For infrequent, but anticipated, highly specialised work the use of contract labour* must be considered.

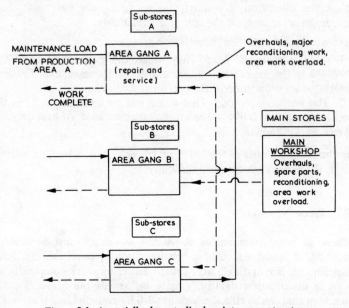

Figure 5.1. A partially decentralised maintenance structure

To summarise, the aim is to balance the advantages of centralisation against the costs associated with travelling; the solution is often a partially decentralised system (see *Figure 5.1*). In this the area maintenance gangs carry out the emergency work as priority and also as much of the scheduled work as possible, while the functions of the main workshop are

1. to act as a reserve of labour for the area gang,
2. to undertake the major reconditioning and overhaul work,
3. to act as a base for those tradesmen who are better centralised, e.g. greasers, inspectors,
4. to co-ordinate (with work-planning) the externally contracted work.

* Some large organisations have seriously considered the possibility of contracting all maintenance work out. Although difficult with existing plant, careful thought is being given to designing new plant with this in mind, the point being that internal labour can now almost be regarded as a *fixed* asset.

5.2.3 Tradeforce Size

If the work load is mainly deterministic it is not difficult to determine the best size of tradeforce. However in most manufacturing situations the work load usually contains a large probabilistic component. Such work occurs with random incidence, has randomly varying completion times, and is usually of high priority. It is best analysed as a complex queueing situation. *Figure 5.2* illustrates a simple repair situation modelled as a simple queue.

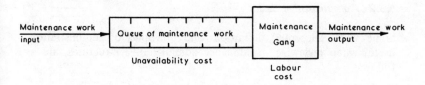

Figure 5.2. The repair situation as a simple queue

The basic aim must be to achieve a tradeforce size and structure which will minimise the total cost of labour and waiting-time (other related decisions concern priority rules, methodisation of work, rate of work, etc. and will be discussed in Section 5.4). If the tradeforce is small, indirect costs as a result of queueing will be high. Conversely, if the tradeforce is large, labour costs will be high. Clearly the critical factor is the cost of unavailability and in situations where this cost is high, low tradeforce utilisation *may* have to be accepted as the optimum solution.

Utilisation can be improved if the structure allows for peak loads (see *Figure 5.7*) to be shared either with a central workshop (see *Figure 5.1*) or with contract labour. The problem of utilisation obviously becomes less difficult the greater the proportion of scheduled work since it is then much easier, with good planning and supervision, to even out the work load.

Operational research techniques such as queueing theory and computer simulation can be used to estimate the optimum size and distribution of the maintenance tradeforce (see Chapter 7). However, *in most cases clear understanding of the maintenance situation coupled with efficient costing will enable a sound resource structure to be developed and subsequently adjusted as necessary.*

5.3　Administrative Structure

5.3.1　Introduction

There is no universally ideal administrative structure for maintenance. Every structure should be based on the principles of administrative theory and should suit the complexity of the maintenance work and the spectrum of resources. That is, the administrative structure should be designed in conjunction with, and to suit the resource structure.

5.3.2　Administrative Basics

Some widely accepted definitions and principles of administration, useful when reviewing maintenance administrative structures, are as follows:

Authority means a power to manage and give orders.
Responsibility is a natural and logical consequence of authority; responsibility and authority should be as nearly equal as possible.
Delegation Authority should be delegated as far down the management line as possible. Responsibility is not shed by delegating work.
Span of control A man can directly supervise only a limited number of people whose work is inter-related. The optimum number depends upon the situation but is between 4 and 8.
Chain of command should be as short as possible. Obviously this is difficult in a large organisation where a more important consideration is the balance between span of control and chain of command.
Unity of command means that each man should have only one boss. Clear divisions of authority are a necessity.

5.3.3　Staff and Line Authority; The Production-Maintenance Relationship

Line executives are those who supervise activities that contribute directly to the profits of the company; staff executives are those who contribute indirectly by providing services or advice to line executives. Staff executives may be said to have staff authority over line executives, in certain cases. A general principle for such situations is that the areas of responsibility and authority must be clearly defined.

An extension of this problem, shown in *Figure 5.3,* is where the staff executive has a specialist team. This structure has the advantages

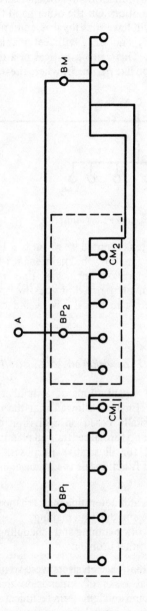

A to C — Level of management
P — Production , suffix indicates plant area
M — Maintenance

Figure 5.3. Centralised maintenance authority

of (a) easy formulation and introduction of new specialist ideas, and (b) clear lines of specialist promotion. On the other hand the foremen, CM_1 and CM_2 of *Figure 5.3*, will have split loyalties, and more importantly production managers, BP_1 and BP_2, will feel unable to control important parts of their work. This can be critical in a decentralised maintenance-production situation like *Figure 5.1* where the unavailability costs are high.

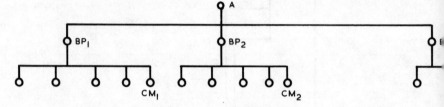

Figure 5.4. Maintenance staff authority

An alternative structure often adopted (e.g. in the U.K. by the CEGB) is shown in *Figure 5.4*. In this, the maintenance foremen are directly responsible to the operation engineers, thus relieving the maintenance engineers of much of their administrative work, leaving them to concentrate on providing technical advice and on long-term engineering planning.

5.3.4   The Effect of Multi-trade Supervision on Administrative Structure

A survey[1] of the maintenance organisations of a number of medium size companies in the N.W. of England showed that the maintenance administrative structure was usually based on a division of the work into specialisations, with little or no formal centralised planning function. A typical structure, simplified for illustrative purposes, is shown in *Figure 5.5*. In theory, the main functions of the manager and engineers were to

1. set maintenance objectives and determine maintenance policy,
2. assist with technical advice and decision making,
3. assist with medium-term work planning and scheduling,
4. look after day-to-day personnel problems.

In most cases it was found that the managers were so involved with the short-term non-professional tasks of supervision that they had little time available for the medium and long-term technical and planning functions. In effect, the managers were being controlled rather than controlling. One company appreciated the difficulty and re-organised their structure as shown in *Figure 5.6*. The staff size was increased by one and the administrative structure divided according to work function rather than work type. The supervisory duties were carried out mainly by the maintenance superintendent, leaving technical advice and long-term maintenance planning to the maintenance engineers. The work

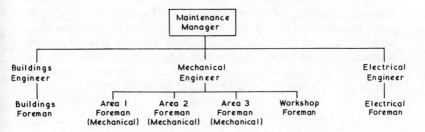

Figure 5.5. Typical maintenance administrative structure, divided according to work type

planning engineers had the responsibility of medium and long-term work planning, also co-ordination of short-term work planning for difficult plant areas. Day to day planning remained the responsibility of the foreman. Where necessary (i.e. area repair, inspection) multi-trade

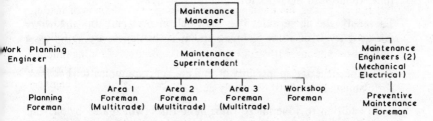

Figure 5.6. Maintenance administrative structure, divided according to work function

gangs were supervised by a single foreman. The advantage of the re-organisation was better utilisation of the qualifications and abilities of the management; this resulted in a better directed department, a consequent reduction in maintenance work and a higher utilisation of the tradeforce.

5.4 Work Planning and Scheduling

5.4.1 Introduction

The previous sections have dealt with ensuring that the right level of resources is available, correctly located and directed in order to meet the maintenance load. This can be regarded as the *statics* of maintenance management. What must now be discussed is the *dynamics* of maintenance management, that is the design of an effective work planning, scheduling and control procedure. In simple terms the function of such a procedure is to get the right resources to the right place, to do the right job in the right way, and at the right time. The objective of such a procedure is to carry out this function at minimum overall cost.

5.4.2 Fundamentals of Maintenance Work Planning

It is essential when designing a maintenance work planning system to observe the basic rules of work control. The most important of these are:

1. The work planner must have the authority (or access to it) to take the necessary decisions (i.e. the allocation of priorities) which affect the work load and the resources.

2. The work planner must have the right information, at the right time, about both the work load and the resources.

To satisfy the above rules it is essential, in most situations, to have different levels of work planning and it is then necessary to follow a third rule, i.e.

3. The areas of responsibility of, and lines of communication between, the planning levels must be clearly defined.

In addition to adhering to these basic rules (which are obvious, but not often followed) it is necessary to have a thorough understanding of the effect of maintenance work load characteristics, as outlined in Section 5.2 on the operation of the maintenance system.

The *probabilistic load* consists of the emergency and corrective maintenance. Its incidence is random, repair times exhibit large variance, and at best only the average load is known. The work load fluctuates as shown in *Figure 5.7* and the repair situation can be represented by a

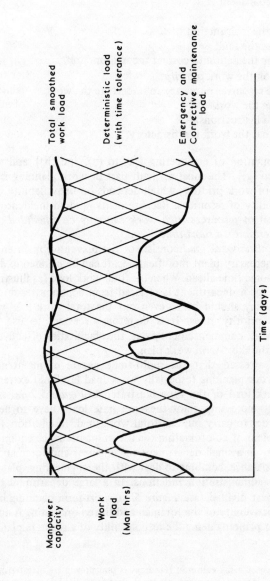

Figure 5.7. The use of deterministic work in work load smoothing

queueing model as in *Figure 5.2*. In such situations the function of the
maintenance department is to

(a) *Locate* the malfunction.
(b) *Diagnose* the fault.
(c) *Prescribe* the action to correct the malfunction.
(d) *Decide* on the work *priority*.
(e) *Plan* the resources necessary to undertake the work.
(f) *Schedule* the work.
(g) *Issue* job instructions.
(h) *Check* that the work is satisfactory.

This is a combination of engineering [(a) to (c), and (h)] and work
planning [(d) to (g)]. The most difficult part of work planning is (d),
the allocation of work priority, which depends on unavailability costs
and the availability of resources. Because of its random incidence and
the limited level of resources *such work can only be scheduled, with
any degree of certainty, a short time ahead.*

The *deterministic work load* consists of the preventive maintenance
work and the necessary plant modifications. It can be planned in detail
and scheduled some time ahead. When the total work load (as illustrated
in *Figure 5.7*) for a department is serviced from common resources*,
the planning policy should be to even out the total work load while
conforming to priority rules. It is therefore necessary to use time
tolerances on work commencement dates, thus facilitating effective job
slotting within the short-term work plan.

It must be stressed that the multi-trade nature of maintenance
work renders work planning (especially work load levelling) extremely
difficult. A work load of the type illustrated in *Figure 5.7* exists for
each trade, but more important these separate loads have to be co-
ordinated in order to carry out the total work load. In addition, there
is also the problem of co-ordinating the use of maintenance equipment
and spare parts. In a small department the engineer and foreman can,
with clerical assistance, combine to deal with the engineering, planning
and day-to-day administration functions. In a large department these
functions are best divided (see *Figure 5.6*), short-term planning being
mainly the responsibility of the foreman and the co-ordinating function
and longer-term planning being the responsibility of a separate planning
department.

---

* The disadvantage of using common resources is that preventive maintenance is
often neglected. Consequently many practical situations divide the man-power,
balancing high utilisation against effective preventive maintenance.

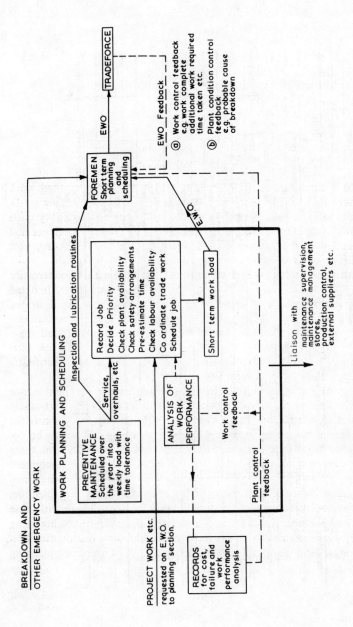

Figure 5.8. Outline of a work planning and scheduling procedure

ENGINEERING WORK REQUEST No.

	Subwork Request Issued				

Requesting Dept.	Requested by:	Extn. No.	Date Issued:	Date Required:

Location No:	Class of Work	Eng. Sec. Ref.	Plant No.	Planning Dates Required		Planning Priority	Plan week No.

Instructions:

MULTI TRADES INVOLVED			
Section	Man. Hr. Slot	Mat'l. Slot	Priority Slot
Building			
Electrical			
Fitting Shop			
Machine Shop			
Welding Shop			
Sheet Metal			
Heavy Gang			
Engraving Shop			
Boiler Shop			
Instruments			
Dept. Fitters			
Installation Dept.			

Authorised by:		Date Completed:	Comp. Sign. Super.	Spaces Opposite for Planning use only.	Total Man Hrs.	Total Material Cost

Figure 5.9. An engineering work order

5.4.3 A Maintenance Work Planning Model

An outline of a work planning procedure for a larger, partly de-centralised, maintenance department (see *Figure 5.1*) is shown in *Figure 5.8*.

Maintenance services are requested for project work, modifications and other such work originating outside the maintenance department, through an engineering work order (e.w.o.), an example of which is shown in *Figure 5.9*. The responsibility of the planning section is to schedule the project work, preventive maintenance, and the necessary corrective maintenance, and to feed this through to the foreman on an e.w.o. and in a form suitable for short term planning. The foremen have the responsibility for the short term planning, and as such the breakdown work goes direct to them. An e.w.o. may be made out by the foreman during or after the breakdown, and is used in this case mainly for work control feedback.

Effective lines of communication are essential to maintenance work planning and control. Close liaison is required between tradeforce and supervision, between supervision and planning, in order to cater for such factors as breakdown priorities, resource shortages, additional work, incomplete work. Close liaison is likewise necessary between the planning section and other plant sections in order to determine the priority procedure for breakdown and other work, plant availability, spares availability etc.

The e.w.o. feedback in conjunction with time cards can provide valuable information for work planning, in particular the planning and scheduling of project and modification work. Analysis of such feedback can determine the average time per period that each trade, in each plant area has available for project work. This average time can then be used as a basis for the planning of projects over several periods. Project planning can be updated at the end of each period and additional resources obtained if necessary.

It has been shown that the planning and scheduling of preventive maintenance work is part of the general planning procedure. It would be useful at this point to discuss in more detail the essentials of preventive maintenance documentation, planning and feedback.

Plant listing and records. These are fundamental requirements of a preventive and general maintenance system. For each unit of plant, classified into units and items and identified according to location and process, it is necessary to enter the following information on a suitable *record card*:

1. Plant information — maker's name, service engineer, essential plant details, spares availability, cross reference to drawing and manual files.

2. Record of preventive maintenance necessary — the type of work, frequency, trades, time, etc.

3. Plant history — major work done, costs, breakdown record and description, remedial action, etc.

Work scheduling. The preventive maintenance load for the year is divided into weekly packages, the objective being to even out the work load. This is usually accomplished with the aid of yearly *planning charts* and, where possible, individual jobs are given a time tolerance. For preventive work of monthly occurrence or less a triggering procedure is required to indicate what work is necessary for any particular week. Triggering devices include the following:

(a) visual checks against a yearly planning chart,
(b) card index files[2], suitable for up to 5000 activities,
(c) sorter-printer machines[3], suitable for up to 80 000 activities,
(d) computer programs[4].

The preventive maintenance work for any one week can be transferred on to an e.w.o., automatically in the case of (c) and (d), and passed on to the foreman for short term planning. Detailed job instructions, *work specification cards* can be prepared for major activities and filed in a location suitable for the tradeforce to use as and when necessary.

Feedback. In addition to the feedback required for work control it is necessary to provide information for the control of plant condition, e.g. description of failure, apparent cause of failure, date of failure, etc. The e.w.o. can be designed to carry this information but in many cases a separate *plant record card* is used. This information is fed back to the planning section to be recorded for future analysis. Because of the difficulties usually encountered in getting the tradeforce to use report forms the information requested must be reduced to bare essentials. A number of preventive maintenance systems recently investigated by the authors have been using supervision to fill in the plant report card.

REFERENCES

1. Holt, D., *A Survey of Maintenance Organisation in the N.W. of England,* B.Sc (Manchester) project report, June (1972)
2. Kalamazoo Ltd.
3. Addressograph — Multigraph Ltd.
4. I.B.M., Maintenance and Engineering Management Information Systems, I.B.M.(UK) Ltd, Feltham, England.

Chapter 6

Quantitative Techniques as an Aid to Maintenance Organisation

There are several well established quantitative techniques which can be helpful when determining the optimum disposition and size of maintenance resources. Of these, queueing theory and simulation will be discussed in this chapter, scientific inventory control in Chapter 7.

6.1 Queueing Theory

It was shown in Section 5.2.3 that the repair situation can be represented by a queueing model, the *simplest* form of which was shown in *Figure 5.2*. It was pointed out that the task of the maintenance manager is to manipulate the factors under his control so as to minimise the sum of the direct and indirect costs of the repair situation. Thus, in theory, he might

1. change the number of repair gangs,
2. alter the queueing discipline,
3. change the repair gang structure,
4. improve the repair rate by improving work methods, motivation or equipment,
5. reduce the input rate by improving preventive maintenance or design-out procedures.

Queueing theory is the mathematical analysis of queueing situations and although it has been used to determine the optimum breakdown gang size[1] it is not sufficiently flexible to reflect the complexities of maintenance. Nevertheless it does illuminate the dynamics of the repair situation.

6.1.1   Fundamentals

Figures 6.1 and *6.2* model the simplest maintenance queueing situations. The jobs arrive at the maintenance department, perhaps wait in a queue, and are dealt with by a repair gang.

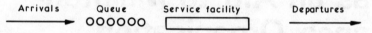

Figure 6.1. A simple single-channel queueing situation

Figure 6.2. A multi-channel queueing situation

A queue forms when the arrival rate exceeds the repair rate and if this is the average condition then the queue is said to be unstable. However, even when the reverse is the case queues can form if the incidence of arrivals and the variation in service times are probabilistic and, for short periods, the arrival rate exceeds the repair rate. Thus in this latter situation the queue length is continually changing but always finite and the queue is said to be stable. Prior to mathematical analysis of a queueing system the following, minimum, information must be obtained:

1. The probability distribution, expressed as an analytical function or p.d.f. (see Section 2.2.2), of the intervals between the arrivals of consecutive jobs. It has been found[2] that the distribution of inter-arrival times for emergency and corrective maintenance jobs is effectively negative exponential, since all times of arrival are equally likely and thus the incidence of arrivals per unit time has a Poissonian distribution.

2. The probability distribution of repair times, expressed as a p.d.f.

3. The queueing discipline, i.e. the rules governing the queues.

4. The repair gang structure (e.g. single repair gang, or multiple gangs in parallel).

6.1.2 Simple and Multi-channel Queueing Models

The *simple queue* (*Figure 6.1*) assumes that both the arrival and service incidence are Poissonian and that the discipline is 'first in, first out' (FIFO) with no restriction on the queue length. Let

Mean arrival rate of jobs   = λ per unit time
Mean service rate of jobs   = μ per unit time

Utilisation factor   = $\rho = \dfrac{\lambda}{\mu}$

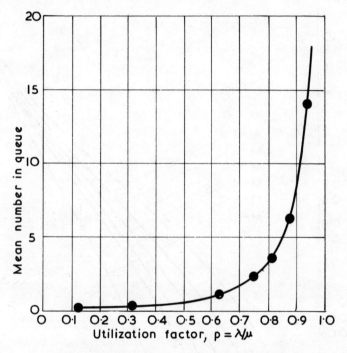

Figure 6.3. Relationship of queue length to the factor ρ

then the equations governing the simple queue can be shown[3] to be

(a) the probability of *n* jobs being in the system*

$$P(n) = \rho^n (1-\rho)$$

* System = queue + repair station.

(b) the average number of jobs in the queue

$$L_q = \lambda^2/\mu(\mu - \lambda)$$

and

(c) the average waiting time for a job in the queue

$$W_q = \rho/(\mu - \lambda)$$

The effect of the utilisation factor, ρ, on queue length is illustrated in *Figure 6.3*.

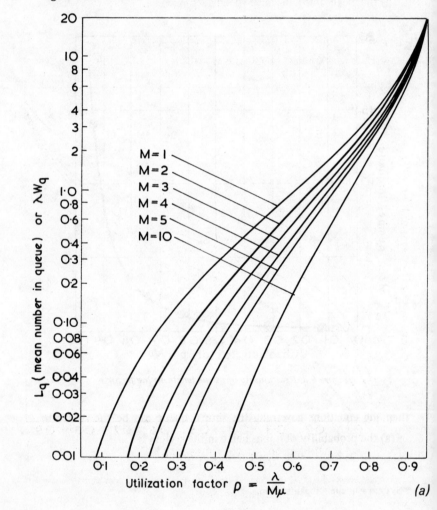

(a)

Example A.   Simple queue.

A factory has many identical pneumatic machines. A study is made of the time between the arrival of machine breakdowns and of the time required for repair. Both distributions are found to be adequately described by the negative exponential p.d.f. The average time between arrivals was found to be 60 min and the average time for repair 50 min.

If repairs are carried out by one gang (three men) calculate

(a) the mean number of machines in the queue,
(b) the mean time a machine spends in a queue,
(c) the utilisation of the gang.

Mean time between arrivals is 60 min, i.e. $\lambda = 1/h$

Mean time of repair is 50 min, i.e. $\mu = 1.2/h$

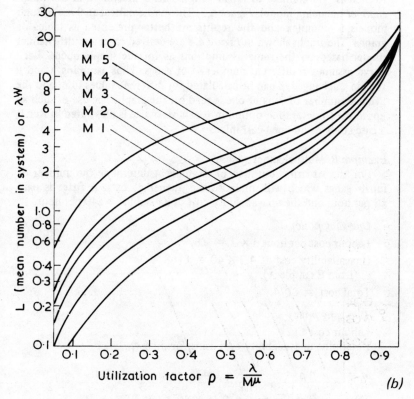

Figure 6.4. (a) Mean number in queue; (b) Mean number in system
(M = number of gangs)

Utilisation factor $\rho = \dfrac{\lambda}{\mu} = \dfrac{1}{1.2} = 0.83$

Mean number of machines $L_q = \dfrac{1}{1.2\,(1.2-1)} = 4.2$

Mean time the machine spends in a queue $W_q = \dfrac{4.2}{1} = 4.2\,\text{h}$

Utilisation $= 1 - P(o)$

$ = 1 - (1 - \rho)$

$ = 1 - (1 - 0.83) = 83\%$

One of the most obvious practical ways of reducing queue length is to increase the number of repair gangs. This is illustrated in the *multi-channel* queueing model, *Figure 6.2*. The mathematical analysis of such models is complex and the results are better presented as tables or graphs. The graphs shown in *Figure 6.4* are derived from a multi-channel model based on the same assumptions as for the simple queue. Each curve assumes a different number M of gangs. Thus assuming λ and μ remain unchanged, ρ can be calculated for different values of M and the average number of jobs in queue read off the graph. The use of such an approach to determine optimum repair staff size is illustrated in Ref.1, a case study of an open-cast mine.

Example B. Multi-channel queues.

For the situation outlined in Example A determine the number of repair gangs which will minimise the queueing costs. A fitter is paid £2 per hour and the unavailability cost of a machine is £40 per hour.

One-gang policy

Labour cost per hour $3 \times 2 = $ £6.

Unavailability cost $= 4.2 \times 40 = $ £168
    (from Example A).

Total cost $= $ £174.

Two-gang policy

Labour cost £12.

Determine unavailability cost using *Figure 6.4(a)*, i.e.

$M = 2$ and $\rho = \dfrac{1}{2 \times 1.2} = 0.415$ giving $L_q \approx 0.18$

Thus unavailability cost $= 0.18 \times 40 = $ £7.20.

Total cost $= $ £19.20.

Three-gang policy

Labour cost £18

$M = 3$ and $\rho = \dfrac{1}{3 \times 1.2} = 0.28$ giving $L_q \approx 0.02$.

Thus unavailability cost $= 0.02 \times 40 = £0.80$.

Total cost $= £18.80$.

i.e. use a three-gang policy.

To the practising maintenance engineer it will be evident that application of queueing theory to maintenance is limited. The maintenance situation is far too complex to be represented by even the most sophisticated queueing model*. Priority rules alone present, in many cases, a serious constraint on effective mathematical modelling.

6.2 Simulation

This is a technique which is sufficiently flexible to model the complexity of the maintenance situation. It is a tool which can be used to find solutions to complex problems involving many variables, some or all of which may be of a probabilistic nature. In effect, the technique permits comparison of alternative courses of action by operating, on a computer, a numerical model of the maintenance system under consideration. It allows different proposals for changing a system to be compared before committing large sums of money.

Simulation has been used in a number of studies to determine the optimum size of maintenance repair gangs. The following example was carried out by an undergraduate at the University of Manchester. For the purposes of illustration it has been somewhat simplified.

A maintenance system served nine identical automatic food processing production lines. The plant operated on an eight hour single shift. Production management suggested that to meet future demands it might be necessary to adopt a three shift per day, five day per week, policy. It was also suggested that every 24 h each line should be stopped for 20 min for essential mechanical adjustments. Taking the nine lines as a whole, these stoppages were to be evenly distributed throughout the day. This new policy would also entail considerable increase in the preventive and corrective maintenance load which had to be carried out at weekends.

* There might be small problems within the maintenance department that might be solved by queueing theory, e.g. stores service.

This proposal for a change in production policy raised the question of maintenance manning policy. One suggestion was to operate a small gang on each shift and to supplement this with weekend contract labour. The shift gang would carry out their duties in the following order of priority:

(a) essential mechanical adjustments,
(b) emergency repairs,
(c) running preventive maintenance.

The problem was to determine the optimum size of shift gang, i.e. that size which would minimise total cost.

After careful study it was decided that in order to simulate this operation the following data were required:

1. The frequency distribution of the time intervals between repair calls for each machine; since each machine is identical one such distribution was sufficient (see *Table 6.1*).
2. The frequency distribution of the maintenance times per machine stoppage.
3. The frequency distribution of the maintenance times per running call, and the distribution of the times to breakdown if no mechanic was available.
4. The ratio of stoppage calls to running calls.
5. Cost factors:
    (a) internal labour rate for week-day and for overtime,
    (b) contract labour weekend rate,
    (c) machine downtime cost.

A representative model was required, but before describing the construction of this there are two concepts, namely *random numbers* and *distributional sampling,* which must be explained.

A succession of numbers is random if there is no systematic relationship whatsoever governing the order in which they occur; given any n successive numbers there is no way of determining anything about the nature of the $(n + 1)^{th}$ number. A large two-digit random number table, for example, will contain all the numbers from 00, 01, 02, etc., up to 99 arranged in a totally arbitrary order, no one number occurring significantly more often than another. Many methods of generating such tables have been used, from dice throwing and roulette-wheel spinning (hence simulation, as described here, often being called *Monte Carlo calculation*) to the sophisticated methods of modern digital computers which usually have a library program for generating random numbers. Tables of numbers, which have been rigorously tested for

Table 6.1. FREQUENCY DISTRIBUTION OF INTER-ARRIVAL TIMES

Repair call inter-arrival times, h

0–1	1–2	2–3	3–4	4–5	5–6	6–7	7–8	8–9	9–10	10–11	11–12	12–13	13–14	14–15	15–16	16–17
Frequency																
1	2	3	6	9	10	12	14	11	9	7	5	4	3	2	1	1

Table 6.2. FREQUENCY DISTRIBUTION IN SERIALISED FORM

Repair call inter-arrival times, h

0–1	1–2	2–3	3–4	4–5	5–6	6–7	7–8	8–9	9–10	10–11	11–12	12–13	13–14	14–15	15–16	16–17
Initial serial number																
00	01	03	06	12	21	31	43	57	68	77	84	89	93	96	98	99

complete randomness, are available in many well known published compilations of statistical and mathematical data.

Distributional sampling can best be described by reference to the histogram shown in *Figure 6.5* which was derived from *Table 6.1*. Note that the items allocated to each class interval in this frequency distribution are numbered serially, in this case from 00 to 99. This information was stored more conveniently in the condensed form shown in *Table 6.2* which lists only the critical points of the histogram. The repair calls that were fed into the model were generated by random sampling from this table. Numbers were drawn, from a table[6] of two-digit random numbers, in a systematic way (every third number, say, working along

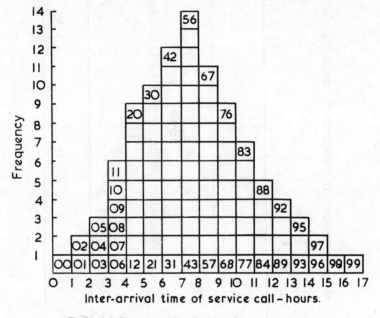

Figure 6.5. Frequency distribution of inter-arrival times

Table 6.3. SAMPLED DATA

Random number	19	06	21	47	89
Next service call, h	4.5	3.5	5.5	7.5	12.5
Cumulative time, h	4.5	8	13.5	21	33.5

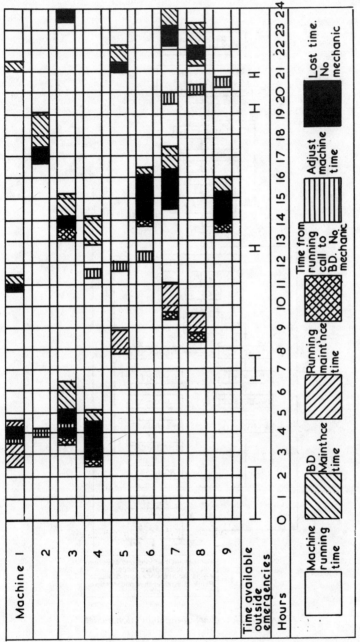

Figure 6.6. Gantt chart for a single-mechanic policy

112

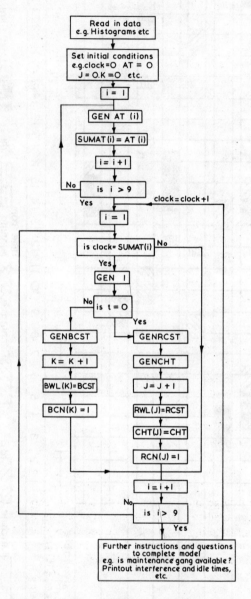

Figure 6.7. Flow chart of the simulation model

the rows or down the columns of the table). If the first number thus drawn was say, 40 then this gave, from *Table 6.2*, a sampled inter-arrival time of 6.5 h. In this way, sampling from *Table 6.2* produced *Table 6.3* for a particular machine. Such a procedure, not involving a computer, is called a *hand simulation.*

A model of the system under investigation was then built and operated by linking together, on a suitable time scale, its various components (the frequency distribution and so on) in a manner which copied the real system. One method of doing this was to follow a representative path through the data, sampling from the frequency distributions and recording the results on a Gantt chart.

In this particular problem the effect of alternative manning policies on machine downtime was of interest. *Figure 6.6* shows part of a one-week hand simulation of the one-mechanic case. It is evident even from this simple example that hand simulation is very time consuming, hence the preference for the use of computers. A flow chart of the simulation model from which a computer program can be written is shown in *Figure 6.7*. It should be stressed that even for the most complex problems a hand simulation, prior to computer programming, is most helpful in determining the logic of the system.

Using the computer model alternative policies were evaluated by starting with a single-mechanic policy and re-iterating with increasing gang sizes. The information required from the computer was as follows:

1. Cost of production idle time due to inadequate maintenance service.

2. Cost of weekday maintenance labour.

3. Extra cost of weekend maintenance labour required to clear work load.

This information was plotted on a graph of total cost versus maintenance gang size thus giving that gang size which satisfied a minimum cost criterion (see *Figure 6.8*).

It must be stressed that simulation will not necessarily give an optimum solution. However, if the model is a realistic one and if the sample size is large enough then a good solution should result. Simulation has been used to solve a number of problems of this type. There are several other primary maintenance problems that simulation could help to solve: preventive versus failure maintenance, internal versus contract labour, repair versus replacement, optimum inspection frequencies, replacement parts inventory, and many more. When considering these problems individually simulation is not always the easiest or cheapest method of solution. Most of them are, however, very much inter-connected and usually cannot be considered separately.

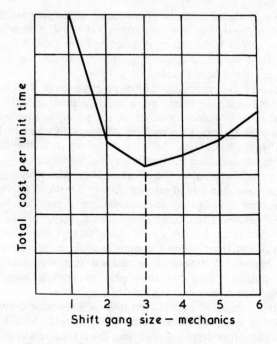

Figure 6.8. Curve of total cost vs. *shift gang size for
nine mechanics in service*

Hence the great advantage of simulation, namely its ability to model
the most detailed and complex problems. If required, the operation of
a whole maintenance department could be simulated in order to examine
the effects of the policies outlined above on the search for a minimum
cost solution.

REFERENCES

1. Jardine, A. K. S. (Ed), *Operational Research in Maintenance,* 119, Manchester
   University Press (1973)
2. Buffa, S. B., *Modern Production Management,* 757, Wiley (1965)
3. Morse, P. M., *Queues, Inventories and Maintenance,* Wiley (1963)
4. Peck, C. G. and Hazelwood, R. N., *Finite Queueing Tables,* Wiley (1958)
5. Wilkinson, R. I., 'Working Curves for Delayed Exponential Calls Served in
   Random Order', *Bell Systems Tech. Jour.,* 32 (1964)

Chapter 7

Spares Inventory Control

7.1 The Objective

In Section 1.3 maintenance was regarded as the operation of a pool of resources directed towards controlling plant availability. These resources were subdivided (see *Figure 1.3*) into spares, equipment and labour, the setting and control of the *level* of which was taken as one of the main factors in the formulation of policy. This chapter deals with the setting and control of the level (i.e. the operation of the inventory) of maintenance spares.

As with other maintenance activities the normal objective of spares inventory control is the minimisation of the sum of the associated costs, direct and indirect. Direct costs could be reduced to zero by totally eliminating spares holding — the consequences for plant availability and hence indirect costs are obvious. Conversely, plant availability would be greatly increased if large numbers of every conceivable spare were always held, but direct costs would then be prohibitive. Seeking the happy medium, the optimum level of spares holding, and economically maintaining it, are the basic objectives of spares inventory control.

7.2 Complicating Factors

Like all other maintenance activities this simply stated task is beset with complicating factors, nearly all of which arise from the variety and complexity of the many thousands of different items, of widely varying cost and usage rate, held by the typical maintenance stores.

Meaningful grouping (e.g. into abrasives, bearings, compressor parts, metals-ferrous, metals-non-ferrous, pipe fittings, valves, etc.), *sub-grouping* (e.g. manufacturer, size, colour, etc.), *nomenclature*

115

(e.g. VALV 1/4 GATE BR 200 SE) and *cataloguing* are a necessary preliminary to any control system. Procedures for dealing with this complex task have been published[1].

Costs and usage rates of the various spares, even for a single plant, usually range over many orders of magnitude. An undergraduate project study, carried out under the supervision of the authors, revealed that the total spares inventory (valued at £1.1M) of a U.K. chemical plant could be divided into three groups, as follows:

(1) *Insurance inventory* (£230 000). Items which were unlikely ever to be used but which were bought as an insurance against failure (£200 000 of these were of individual value greater than £500, were capitalised and held by plant management. Only the £30 000 of smaller items were attributed to spares inventory).

(2) *Manually controlled inventory* (£860 000). Items of such low usage rate and hence large percentage variation and poor predictability that computer analyses, based on the sort of mathematical approach outlined in the next section, could not be used to determine replenishment order quantities and times.

(3) *Automatically controlled inventory* (£10 000). Items of such high usage rate and hence small percentage variation and good predictability that computerised forecasting and order determination could be used.

As in much other modern plant it was felt that the size of group (1) was justified by the high cost consequence, to a high throughput flow process, of such stock-out. The holding of such items on an international, company, basis had been tried, without much success.

Due to the small value of the items involved, it was questionable whether computerised control was at all appropriate in the above case. However, at any given plant the proportion of group (3) type items, as opposed to the group (2) type, can be increased for existing equipment by studying the possibility of interchangeability of spare parts. In the above chemical plant scope for this existed in the pump spares inventory, worth £150 000, since pump technology is now very advanced and designs very similar. Interchangeability can also be extended by stocking only the largest size of a not too widely varying class of spare and reducing, in the workshop, the appropriate dimension to the size required. In most modern techniques, however, plant specialisation is very advanced and the opportunities for interchangeability limited.

Interchangeability can be built in *at the design stage* by modularisation[2], i.e. the integration of a sub-system of several components into one standardised and rapidly replaceable item or 'module'. If any one component fails the whole module is replaced. The standardisation increases interchangeability and the modularisation reduces the number of different items in the spares inventory.

Interchangeability and modularisation multiply the effective number of items in — and the usage rate of — a given spares group or sub-group. It is a statistical corollary of this that the fractional variability (s.d./mean) is reduced and hence the predictability increased.

Long and variable delivery dates, problems in the assignment of ordering decisions to the correct management level, the possibility of re-conditioning some spares in the maintenance workshop, difficulties in the provision and location of proper storage facilities, rapid obsolescence and updating of plant, are other factors complicating the spares inventory problem. The experienced reader will be aware of many more.

Clearly, neat and tidy mathematical analysis is strictly applicable to the control of only a narrow range of high usage spare parts. Nevertheless, such analysis does constitute the starting point of any rational approach and the remainder of this chapter will be devoted to it.

7.3 Scientific Inventory Control

The problem is to balance the cost of holding stock against the cost of running out. More generally, inventory control theory attempts to determine those procedures which will minimise the sum of the costs of

1. Running out of stock (production loss due to stoppage, cost of temporary hire, etc).
2. Replenishing stock (per item, an inverse function of the order quantity, R).
3. Holding stock (interest on capital, insurance, depreciation; over a wide range of stock levels wages, light, heat, storage facilities, rent and rates can be taken as fixed and therefore not influencing the optimisation analysis).

7.3.1 Simple Model for Evaluating EOQ

F. W. Harris, working at Westinghouse, developed an equation for determining the Economic Order Quantity (EOQ), i.e. that size of replenishment order which minimises the sum of holding and re-ordering costs. The most simplified treatment is as below.

Stock is re-ordered at regular intervals, the lead time L between order and delivery is negligible, stock is not allowed to go negative, demand is assumed to be constant at a rate q per unit time. This highly simplified model is illustrated in *Figure 7.1*.

Let Replenishment quantity = R
 Ordering cost = a
 Holding cost per item of stock per unit time = b
 Total expected cost per unit time = C
Then average stock (see *Figure 7.1*) = $R/2$
so average holding cost per unit time = $Rb/2$
Number of replenishment orders per unit time = qR
so replenishment cost per unit time = qa/R

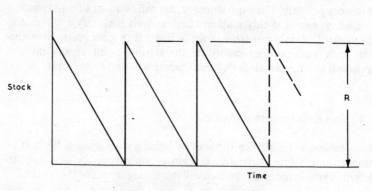

Figure 7.1. Model for evaluation of EOQ

Total cost per unit time is therefore $C = (qa/R) + (Rb/2)$

We wish to find R at which C is a minimum, i.e. where

$$dC/dR = -(qa/R^2) + (b/2) = 0,$$

i.e. at $R = (2qa/b)^{1/2}$

which is the EOQ in this case.

Example.

q = 150 items/year
b = £12/item/year
a = £2/order

Table 7.1. EVALUATION OF TOTAL COST

	Costs (£) for one year		
Replenishment quantity, R	Replenishment cost, qa/R	Holding cost, Rb/2	Total cost, C = (qa/R) + (Rb/2)
1	300.0	6	306.0
3	100.0	18	118.0
6	50.0	36	86.0
7	42.9	42	84.9
8	37.5	48	85.5
10	30.0	60	90.0
20	15.0	120	135.0

Plotting this data we see, from *Figure 7.2*, that the EOQ is approximately 7 units. Using the analytical expression we find

$$EOQ = (2qa/b)^{1/2} = (2 \times 150 \times 2/12)^{1/2} = 7.07 \text{ items}$$

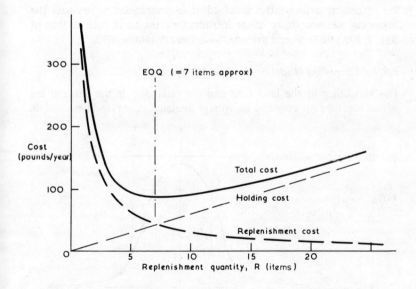

Figure 7.2. Effect of replenishment quantity on costs (data of Table 9.1*)*

7.4   Safety Stock

The above is as simplified a model as can be usefully conceived. It does not take into account those factors, outlined in Section 9.2 which are usually of overriding importance in real situations. Random variability of demand and lead time are two such factors. The consequent risk of running out of stock is mitigated by holding safety or buffer stock. This is considered as a static reservoir not involved in normal turnover, although to avoid deterioration the items in it are, from time to time, used to meet demand and immediately replaced from ordinary stock.

The so-called 'two-bin system' is one method of providing safety stock. A fixed replenishment order is placed whenever stock reaches a pre-set re-order level, r, (e.g. storage in two bins, replenishment order when first bin empty, service from second bin until order received). The value of r depends upon

- (a) the rate of demand and its variability,
- (b) the lead time and its variability,
- (c) the cost of stock out.

Since random variability must be taken into account, mathematical analysis of this system is considerably more complex than that of the simple EOQ model, Section 7.3.1. The reader who is not familiar with the statistical mathematics involved is recommended to by-pass the following section (study of an introductory textbook such as that of Ref. 2, Chapter 2, would provide the necessary familiarity).

7.4.1   Simplified Model of the Two-bin System

The variability in the lead time and the variability in the demand are added together by assigning an average demand, μ, per lead time and an

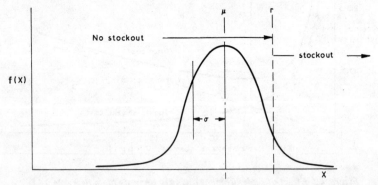

Figure 7.3. Probability density distribution of X, *the demand-per-lead-time*

associated variance, σ^2, in that demand (i.e. it is assumed that a variance can be calculated, or measured, which will be compounded from two sources of variability, uncertainty in lead time and uncertainty in demand). Furthermore, it is assumed that (as illustrated in *Figure 7.3*) demand-per-lead-time is normally distributed (which is reasonable if (a) demand is random and, on average, high and lead time long and (b) the lead time is itself normally distributed).

In addition, it is assumed that the consequences of stockout can be costed precisely, that such cost is extremely high, and that the buffer stock level will consequently be so high that stockouts will occur only rarely.

Let q = average number of demands per unit time
R = replenishment quantity
r = re-order level (capacity of second bin)
s = buffer level (stock in hand just prior to replenishment)
X = demand per lead time
μ = average demand per lead time
σ^2 = variance in demand-per-lead-time
a = ordering cost
b = holding cost per item of stock per unit time
c = stock-out cost per unmet demand

The behaviour of the stock level is illustrated in *Figure 7.4*.

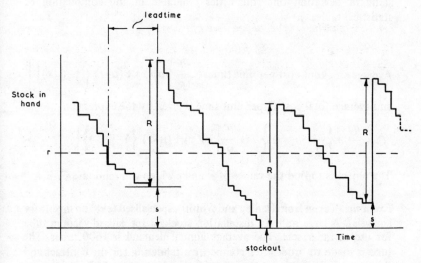

Figure 7.4. Pattern of stock changes, two-bin system

Average value of buffer level $s = \int_0^r (r - X)f(X)\mathrm{d}X$

But if $(r-\mu) >$ two or three times σ then

$$s \simeq \int_0^\infty (r - X)f(X)\mathrm{d}X = r - \mu$$

and average holding cost per unit time $= (\dfrac{R}{2} + r - \mu)b.$

Average number of unmet demands per lead time

$$= \int_r^\infty (X - r)f(X)\mathrm{d}X$$

$$= \int_r^\infty X f(X)\mathrm{d}X - r \int_r^\infty f(X)\mathrm{d}X$$

$$= \mu - \int_{-\infty}^r X f(X)\mathrm{d}X - r \left\{ 1 - \int_{-\infty}^r f(X)\mathrm{d}X \right\}$$

$$= \sigma\phi(x) + (\mu-r) \left\{ 1-\Phi(x) \right\}$$

where $x = (r-\mu)/\sigma$, and $\phi(x)$ and $\Phi(x)$ are the probability density and c.d.f. at value x of a normally distributed variate x of mean zero and standard deviation one (quantities evaluated in any compilation of statistical tables[5]).

So,

Average stockout cost per unit time $= \dfrac{qc}{R} \left[\sigma\phi(x) + (\mu-r) \left\{ 1-\Phi(x) \right\} \right]$

and average total cost C, per unit time is given by the expression

$$C = \frac{qa}{R} + \left\{ \frac{R}{2} + r - \mu \right\} b + \frac{qc}{R} \left[\sigma\phi(x) + (\mu-r) \left\{ 1-\Phi(x) \right\} \right]$$

The object is to find the values of R and r which will minimise C.

Example. (Taken from Hadley and Whitin's specialised text[3] on inventory analysis). A large military installation stocks a specialised vacuum tube for use in radar sets. The average annual demand is 1600 tubes. The tube is made to order after competitive tendering for the contract and the cost of placing an order is $4000. Holding cost is $10 per tube per year. Average demand per lead time is 750 tubes with a standard deviation of 50. If stockout occurs the demand *is* met, at a cost of $2000 per tube, by an emergency ordering and delivering procedure.

The first step is to evaluate the square-bracketed quantity, the average number of unmet demands per lead time, which is a function of r.

Table 7.2. EVALUATION OF AVERAGE UNMET DEMANDS PER LEAD-TIME

r, (tubes)	x $(=\frac{r-\mu}{\sigma})$	$\phi(x)$	$\sigma\phi(x)$	$\Phi(x)$	$[\sigma\phi + (\mu-r)(1-\Phi)]$
850	2.00	0.0540	2.700	0.977 25	0.4250
870	2.40	0.0224	1.120	0.991 80	0.1360
880	2.60	0.0136	0.680	0.995 34	0.0742
890	2.80	0.0079	0.395	0.997 44	0.0366
900	3.00	0.0044	0.220	0.998 65	0.0175
920	3.40	0.0012	0.060	0.999 66	0.0022

We can now evaluate C for various combinations of R and r (see *Table 7.3*).

Table 7.3. EVALUATION OF TOTAL COST ($R = 900$) (All costs in dollars)

r, (tubes)	Ordering cost, qa/R	Holding cost, $(R/2+r-\mu)b$	Stockout cost, $qc[\,]/R$	Total cost per year, C
850	7111	5500	1511	14 122
870	7111	5700	484	13 295
880	7111	5800	264	13 175
890	7111	5900	130	13 141
900	7111	6000	62	13 173
920	7111	6200	8	13 319

This gives a family of curves, as shown in *Figure 7.5*, which indicates a minimum at about $R = 1100$, $r = 890$ tubes. Repeating the process with smaller increments of R, in this vicinity, locates this minimum at $R = 1145$, $r = 890$ tubes. An initial approximation to the result for R can be obtained from the simple EOQ formula, i.e.

$$R = (2qa/b)^{1/2} = 1131 \text{ tubes}$$

showing that when stockouts are rare it is the balance of holding and ordering costs which largely determines R.

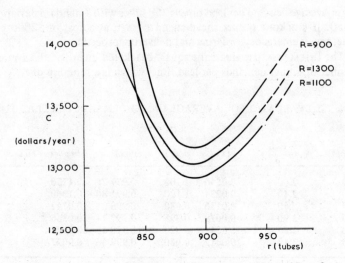

Figure 7.5. Graphical determination of optimum values of R *and* r.
(Data as from Table 9.3*)*

Note that the average buffer, or safety, stock $s = 890 - 750 = 140$ tubes.

7.4.2   The Periodic Review System

In the two-bin system a fixed quantity is ordered at variable intervals of time; in general to operate such a method needs continual monitoring of all stock changes and it is only with the advent of the computer that it has become at all widely used. The more traditional and popular method is the periodic review. At fixed intervals of time (e.g. end of the season, supplier's regular delivery time) the inventory is reviewed and stock ordered in quantities decided as a result of the review.

The simplified two-bin analysis has illustrated the general nature of inventory analysis when applied to a statistical situation. Analyses of periodic review policies, although differing in mathematical detail, are broadly similar. A very full description and review of a wide variety of analytical models is given by Hadley and Whitin[3]. Whitin also claims to have published the first book[4] in English dealing with stochastic inventory analysis.

The two-bin system will, in general, lead to smaller buffer stocks, and hence lower cost, than the periodic review system. In the latter, buffer stocks must offer protection for the lead time plus the period while in the former only the lead time has to be covered. Despite this it is claimed that, in practice, periodic review is cheaper because

1. It allows simultaneous ordering of different items.
2. It minimises day-to-day fluctuations in control and order work load.
3. It is suitable for seasonable items.

A suitable strategy for a large establishment might be to use a combination of systems, e.g. the two-bin system for high value items of fairly constant average demand, dynamic programming[3] for high value items of fluctuating average demand, periodic review at optimum frequency for medium value items, annual review for low value items.

In many practical cases the random variables may be many and may not be reasonably represented by one of the analytical probability distributions. Computer simulation, along the lines indicated in Section 6.3 can then be used, and is well suited to inventory problems.

Readers especially interested in Inventory Control are recommended to consult C. D. Lewis's concise, but very readable, introductory text[7].

REFERENCES

1. Wolf, T. R., *Improving the Efficiency of Maintenance Stores,* In. Com. Tec. Ltd., June (1967)
2. Swärd, K., 'Some Practical Views on the Maintainability Concept', *UNIDO Course on Maintenance Management,* Stockholm (1975)
3. Hadley, G. and Whitin, T. M., *Analysis of Inventory Systems,* Prentice-Hall (1963)
4. Whitin, T. M., *The Theory of Inventory Management,* Princeton University Press (1953)
5. Lindley, D. V. and Miller, J. C. P., *Cambridge Elementary Statistical Tables,* C.U.P. (1952)
6. Kelly, A. and Harris, M. J., 'Simulation, an Aid to Maintenance Decisions', *The Plant Engineer,* **15,** 11, 43, Nov. (1971)
7. Lewis C. D., *Scientific Inventory Control,* Newnes-Butterworths (1976)

Chapter 8*

Network Analysis for the Planning and Control of Maintenance Work

8.1 Introduction

Generally speaking, project work and anticipated shutdowns, which are essentially deterministic work, should be blended with the probabilistic work load, wherever possible. However, there are situations, such as annual overhauls or major shutdowns of petrochemical plant, where such work must be treated as a separate planning problem because of its magnitude, complexity or high unavailability cost. *Barchart display* and *network analysis* are valuable techniques, widely used, for planning and, in particular, *controlling* such shutdowns.

A typical barchart is shown in *Figure 8.1*. It is usually mounted on a wall board with the bars as moveable strips which can be adjusted when determining resource allocation. This technique is quite adequate for small straightforward jobs of no more than, say, 300 activities, but for a major shutdown 3000 to 4000 activities may need to be included for adequate control and this creates problems. It is extremely difficult, for example, to ascertain the logistics of a hundred simultaneous activities or twenty different resource loadings when using barcharts only. Furthermore, the planning system must be capable of quick and accurate response as constraints are altered. The only means currently available for meeting such requirements is *computerised network planning†*.

* Contributed by H. Moody, Senior Project Engineer, ICI (Organics Div.) Ltd., Blackley, Manchester.
† *Manually operated* networks can be useful for smaller shutdowns.

126

Initially, network methods were used primarily to minimise time but later developments have also taken costs and resource utilisation into the overall optimisation. These will be discussed in some detail.

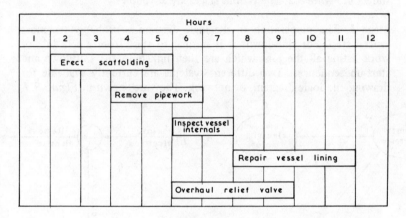

Figure 8.1. Barchart

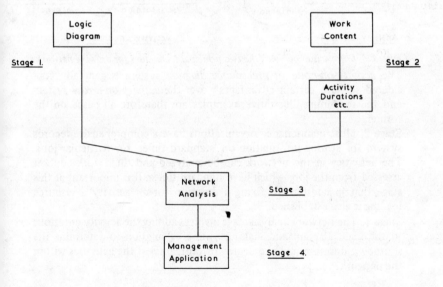

Figure 8.2. System of planning using network analysis

8.2   The Basics of Network Planning

The overall procedure, for the case of a maintenance shutdown, is shown in *Figure 8.2*. The various stages are as follows.

Stage 1. The preparation of the logic diagram is probably the most crucial step. For shutdowns it is normal to start from a schedule of work listing all the jobs which are then linked to the shutdown and start-up sequences. Two different systems are currently available for drawing the logic diagram, or network. They are shown in *Figure 8.3*.

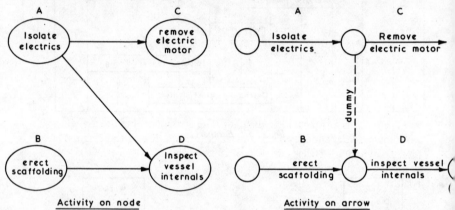

Figure 8.3. Comparison between the two principal techniques for drawing networks

The *activity-on-node* or *precedence diagram* system is generally considered to have certain advantages* over the *activity-on-arrow* system and the remaining illustrative examples are therefore all based on the former.

Stage 2. Most maintenance organisations have a comprehensive records system for storing information on 'standard times' for particular jobs. The activities in the network can be analysed and their work content assessed from the jobs which have standard times. It is important at this stage that in addition to fixing a duration for an activity a resource loading is also established.

Stage 3. The network analysis stage involves adding the activity durations to the logic diagram and making a time calculation to determine the shutdown duration and the degree of criticality of the activities within the network.

* For example, simplicity, ease of modification, easier representation of overlapping activities, etc.

Figure 8.4 shows the basic node. Every activity has two possible start dates and finish dates, the earliest and latest in each case. The degree of criticality of any particular activity is defined as the *total float* (TF), which is simply the difference between the latest and the earliest start. In the event of this difference being zero then the activity is critical.

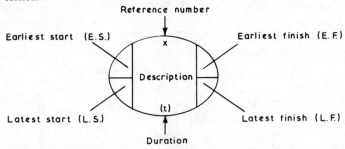

Figure 8.4. The basic node

Figure 8.5 is a typical logic diagram. The arrows joining the activities are known as dependency lines and the interpretation of the diagram is that when activity 101 is complete then activities 102, 103 and 105 can begin; similarly activity 106 cannot begin until activities 102, 104 and 105 are complete.

Figure 8.6 shows the time calculations and the total float for the various activities. The first part of the calculation is to work through the network from left to right filling in the top boxes, i.e. ES and EF. Note particularly that activity 106 cannot begin until activities 102, 104 and 105, i.e. time unit 17, are all complete. The bottom boxes are filled in working right to left through the network so that the criticality of the various activities can be established. In the case of activity 101 there are three dependency lines running back from activities 102, 103 and 105; the lowest value of their latest starts, i.e. 7, is taken as the latest finish of activity 101.

The last part of the calculation is to establish the criticality of the various activities in the network and this is done by subtracting the earliest starts from the latest starts.

The critical path is defined as that chain of activities in the network which has the least total float, i.e. in the case of *Figure 8.6* activities 100, 101, 102 and 106. The critical path establishes those activities which must start at their earliest start and finish at their earliest finish if the shutdown is to be completed on time. Further, as a first step to reducing the shutdown duration it is the critical path activities which require shortening.

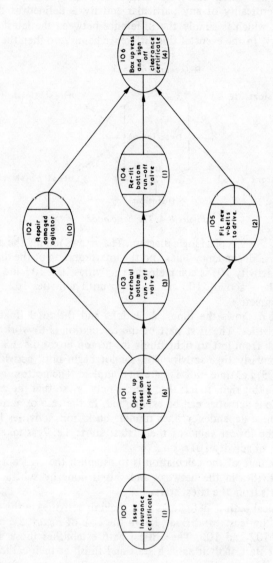

Figure 8.5. *Logic diagram with activity durations added*

131

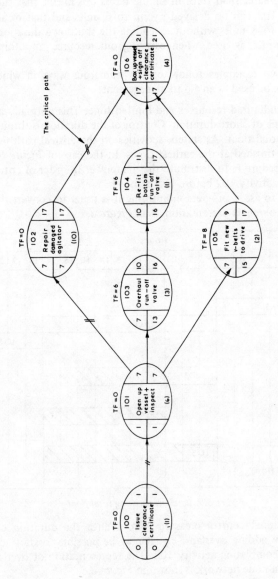

Figure 8.6. Time analysis and identification of critical path

Note that activities 103, 104 and 105 in the above network have total float greater than zero. In simple terms this means that the start of activity 103 can be delayed by up to 6 units and that of activity 105 by up to 8 units without extending the shutdown duration. It is this float which is used when carrying out resource smoothing and levelling.

Stage 4. Two examples follow of the numerous ways in which the network can be used as an aid to management.

Case A – Unlimited resources and limited time. This normally applies to shutdowns of short duration. Control of the duration is through the use of the total float. As stated, activities on the critical path must be started and finished at the earliest times. In the case of *Figure 8.6* the correct procedure is to distribute the manpower in order of criticality, i.e. man up activity 102 before 103 before 105, etc.

In order to assist progress monitoring it is usual to convert the network into a barchart presentation, as in *Figure 8.7*.

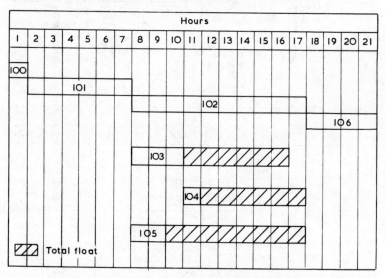

Figure 8.7. Barchart showing total float

An additional control measure is to reduce the duration of the shutdown by adding overlaps, e.g. it may be possible to start activity 106 before completing activity 102. The representation of overlaps in an activity-on-node network is shown in *Figure 8.8*.

The overall effect of introducing the overlap between activities 102 and 106 is not only to shorten the critical path on *Figure 8.6* but also

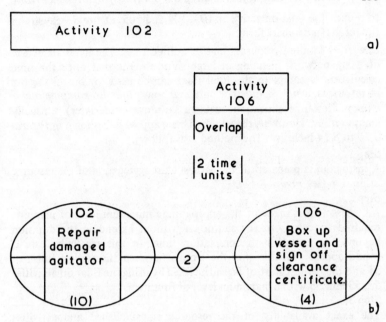

Figure 8.8. Overlapping activities. (a) Overlapping in barchart; (b) overlapping on network

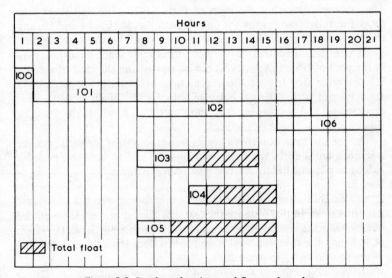

Figure 8.9. Barchart showing total float and overlap

to reduce the float on activities 103, 104 and 105. *Figure 8.9* illustrates the results in barchart form.

Case B — Limited resources and/or limited time. The main attraction of using network planning of large shutdowns is that once the time calculation is complete the total float can be used to control the use of resources, *which include not only manpower but also equipment and money*. Resource allocation on large shutdowns (see later) is usually computerised. However, the standard hand approach is shown in *Figures 8.10* to *8.14* inclusive. The method is as follows.

Step 1.

A histogram is prepared of the earliest start aggregation of the resource utilisation (see *Figure 8.11*).

Step 2.

An assessment is made of the approximate minimum level of manning required to carry out the shutdown without extending the duration beyond the critical path completion time. In this particular case it would appear probable that the trough between the earliest finish of 51 and the earliest start of 56 can be filled by using the float on activities 52, 53 and 55 with a maximum level of four men.

Step 3.

The exact availability of the resource is established and activities identified which can be split between different shifts (see *Figure 8.12*).

Step 4.

The resource allocation is shown in *Figure 8.13*. The first stage is to show the activities forming the critical path, i.e. 50, 51, 54 and 56, and the asumption is made that activity 56 can be split between two shifts.

At time unit 3* activity 51 requires three men which leaves one man in the pool of labour. Activity 53 can be started and the start of activity 52 delayed. Similarly at the completion of activity 53 activity 55 should begin with a further delay to activity 52. Finally activity 56 is commenced and is completed by time unit 15, i.e. within the available float.

One of the main objectives of this step is to obtain for each resource a smooth build up to the peak working level and to hold this peak for as long as possible before starting a smooth run down.

Step 5.

It may be that after the initial resource calculation several activities are left with float. In this instance the resource allocation should be carried out by reducing the manning availability. *Figure 8.14* shows the effect of reducing the manning level from four to three men, i.e. an extension of the shutdown by six time units.

---

* N.B. The whole period is divided, for the purpose of the analysis, into units of four hours.

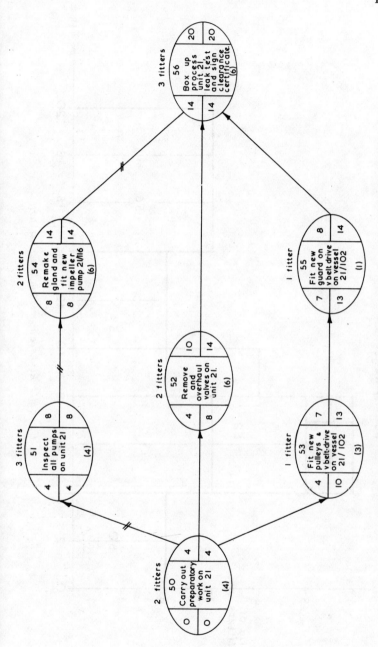

Figure 8.10. Network showing activity durations and associated resource loadings

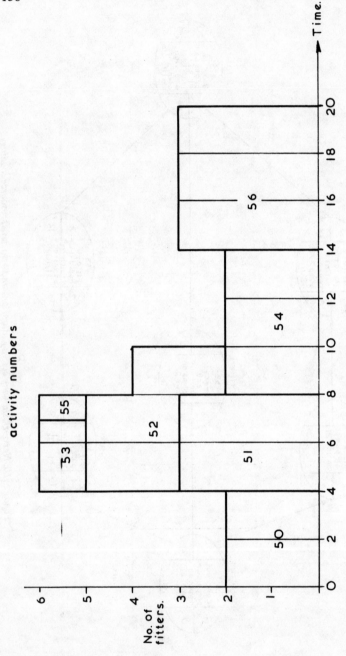

Figure 8.11. Resource aggregation by earliest start

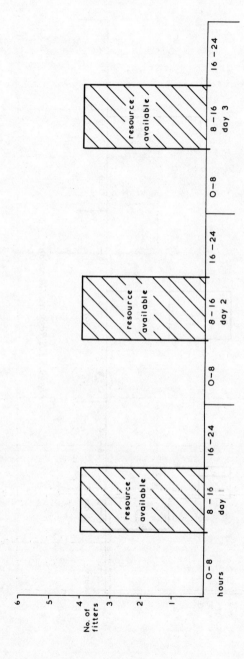

Figure 8.12. Resource availability

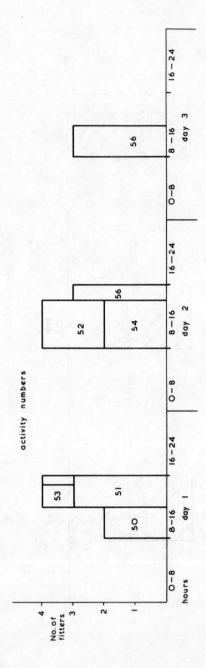

Figure 8.13. Resource allocation

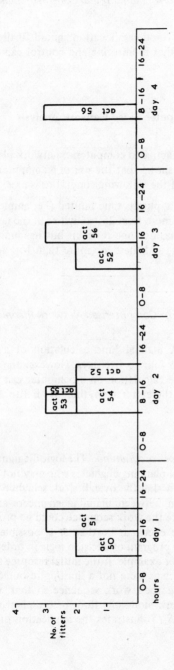

Figure 8.14. Resource allocation – resource limited

Step 6.
The histogram showing resource allocation should finally be translated into barchart form so that monitoring and control can be carried out during the shutdown.

8.3   The Case for Computer Aided Network Analysis

As stated, it is normal practice to computerise network planning of large shutdowns. Experience shows that the use of a computer is justified if, broadly speaking, one of the following conditions is met:

1. If the shutdown is purely time limited (i.e. ample resources are available), and there are more than 250 activities in the network.

2. If the shutdown has no time restraint but has limitations on the availability of resources, and there are more than 150 activities in the network.

8.3.1   The Advantages of Computerised Network Planning

Speed and accuracy   A manual time calculation of a 3000 activity network can be done in four to five days. However the probability of errors is high. The same calculation on a computer can be carried out initially in two days and thereafter in anything from 3 to 24 h depending on the degree of priority.

Prior investigation of work programme   The logic diagrams are normally prepared by professional planning engineers who may not be fully aware of the possibilities of revising the overall work schedule to shorten the shutdown. Consultation between planning engineers and supervisors can highlight areas where the work sequence could be profitably altered and this can readily be checked. Further, it is possible to investigate the introduction to the program of decision logic in order to streamline the use of resources. For example, if the initial resource allocation runs showed that the electricians were not a limiting resource it might then be possible to optimise their work sequence so that, when isolating electric drives, they work along one floor of the plant then along the next floor, etc. *Figure 8.15* illustrates the application of decision logic to a typical network.

Identification of criticality Clearly, the appropriate personnel should be fully aware of the truly critical activities and resource loadings. In the initial calculations activity durations and resource loadings are based on standard times, etc. achieved in previous shutdowns. Key activities can therefore be shortened by, for example, adding more resources. Computer output should be standardised as far as possible when several shutdowns are planned for a particular plant; this assists communication and can lead to greater control over critical areas of the shutdowns.

Optimisation of resource utilisation The main benefit of network analysis of shutdowns is in facilitating the control of large numbers of resources. This is particularly so where contract labour is being used and forecasted requirements must match actual numbers used. On large shutdowns up to 200 contract men can be employed and the planning should be in such detail that, for example, when a fitter has to open a pump for inspection then before he receives the job card the necessary prior activities (disconnection of electrics, issue of protective clothing, provision of essential spares, etc) should be known to be complete.

Issue of useful computer output One of the advantages of the recent developments in computer language is that it is relatively easy to convert standard computer output into any required format. In addition, the output can be sub-divided in detail so that fewer sheets of information have to be issued to cover any one particular case (say, activities involving one class of tradesman on one type of process unit in one area of the plant).

Before detailed shutdown planning goes beyond the logic diagram stage the output required by each level of staff should be established.

Updating during shutdown Any planning system must be capable of coping with changes as and when they occur so that planning is always on the basis of the most up-to-date information of the situation. Experience has shown that the initial stages (production run-down, cleaning out process lines, etc.) of a two- to three-week shutdown usually go reasonably according to programme. The first major review of progress should take place after this first phase and should examine the additional work arising due to unforeseen circumstances. If a decision is made to update the computer program then the revised print-outs should be

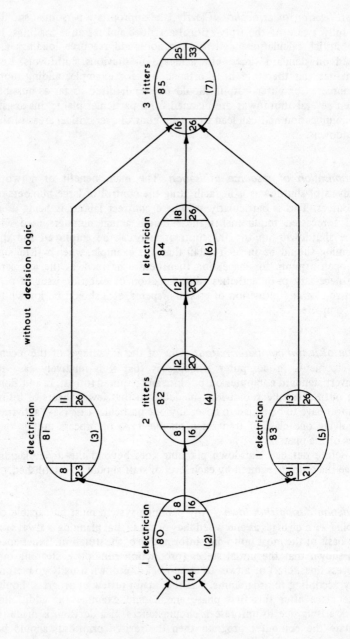

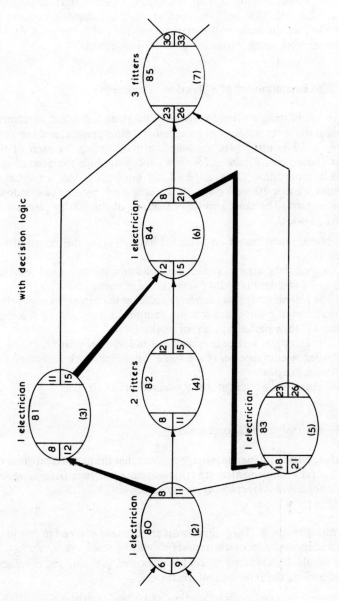

Figure 8.15. The application of decision logic to smooth resource working

available within 5–7 h, i.e. within a normal working shift. During the period when the new print-outs are awaited the shutdown should be controlled on the basis of the existing program, as far as possible, the additional work being 'manually' planned on barcharts.

8.4   The Establishment of a Shutdown Programme

Before establishing a formal system for planning individual shutdowns it is essential that management accepts that it must prepare, and regularly review, a 12-monthly plan showing the basic timing for each of the major shutdowns. *Figure 8.16* shows the procedure for producing a detailed programme. This is based on controlling the sequence of activities starting 10 weeks, i.e. 50 working days, before the shutdown is due to start. The most important features of the 50-day programme are as follows:

1. Management meets on a formal basis to agree the extent of the shutdown.

2. A good data retrieval system, for estimating job times and resource loadings for standard repetitive activities, is in operation.

3. The planning engineers actively encourage the personnel responsible for implementing the shutdown programme to comment on the logic diagram for their particular areas of work.

4. The computer service is available as and when required.

5. Adequate arrangements are made for issuing work schedules and monitoring progress.

6. Preparatory work is properly planned and controlled.

8.4.1   Control during Shutdown

The three basic activities necessary to ensure that the detailed programme is being fully implemented are (i) *monitoring progress,* (ii) *issuing job instructions,* and (iii) *reviewing and updating.*

Monitoring progress   Large shutdowns are normally planned in one hour time units; ideally, progress reports should be made every two hours. They should be delivered to centrally located planning engineers and should include the following information:

(a) the estimated completion time of all 'live' activities,

(b) the actual completion time of activities,

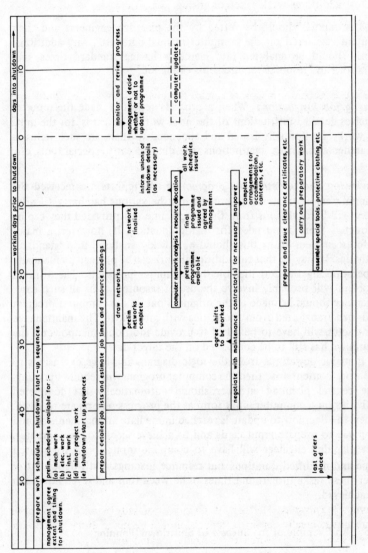

Figure 8.16. Sequence of work required to be carried out to produce realistic shutdown programme

(c) a review of impending activities,
(d) excesses or shortfalls on resources,
(e) additional work arisen or arising.

The reports should be vetted by the planning engineers and their content registered on the computer/manual barcharts. Any additional work should be analysed into elements having standard times, and durations and resource requirements estimated.

Issuing job instructions When a progress report is made the reporter receives detailed instructions of the next work. Job cards, for the units of work, are normally issued and appended to them are such items as clearance certificates, instructions regarding access to special tools, etc.

Reviewing and updating the programme As reports are received the planning engineer monitors progress on the various barcharts. Clearly, there will be deviations from the programme but provided they can be corrected there is no need for major updating. If, however, a major unforeseen item arises this should be quickly analysed into 'standard' activities. It may be that the additional work can be planned by manually superimposing the activities on the computer printouts. The difficult problem will probably involve the use of resources — in all such cases reference should be made to the original float on the computer program and men transferred from those jobs with most float. The maintenance contractor will have to be asked to provide additional manpower if the shutdown has still to be completed on the target date.

If major deviations from the logic diagram and large extensions to the work content arise then the computer program will require updating. The central planning engineer should automatically list completed activities on a computer input form as the progress reports are received. When the decision to update is reached immediate arrangements should be made to prepare input cards and to achieve access to the computer. The planning engineer will have to complete input forms for revised logic and modified durations and resource loadings. With tight control over these tasks turnaround times of between four and seven hours can be achieved.

8.5 An Example of the Success of Shutdown Planning

Between 1969 and 1973 a large chemical plant in the U.K. planned and controlled 25 shutdowns using the system described. Of these shutdowns 20 were judged successful. The most common failures were due to (a) trying to shorten the '50 day' programme (see *Figure 8.16*), and (b) extension of the work content due to unforeseen circumstances. The shutdown season tended to fall into the second quarter of each year thus increasing the probability of shutdowns on different lines occurring simultaneously. At one time three shutdowns were occurring in parallel and some 400 contract men were employed. Nevertheless, two out of the three shutdowns were successfully completed on time and within the planned resource loadings.

One spin-off obtained from computerised network planning was the saving in maintenance planning effort. On a purely manual system some nine man-months were required to prepare barcharts but when using the computer only three man-months were required and this included an allowance for looking at alternative programmes and fully optimising labour requirements.

Chapter 9*

Condition Based Maintenance

Maintenance carried out in response to a significant deterioration in a unit as indicated by a change in a monitored parameter of the unit condition or performance, is called *condition based maintenance*.

Condition based maintenance differs from both failure maintenance and fixed-time replacement in that it requires monitoring of some condition-indicating parameter of the unit being maintained. This contrasts with failure maintenance, which implies that no successful condition monitoring is undertaken and with fixed-time replacement which is based on statistical failure data for a type of unit. In general, therefore, condition based maintenance will be more efficient and adaptable than either of the other maintenance actions. On indication of deterioration, the unit can be scheduled for shutdown at a time chosen in advance of failure, yet, if the production policy dictates, the unit can be left to run to failure. Alternatively the amount of unnecessary preventive replacement can be reduced, while if the consequences of failure are sufficiently dire, the condition monitoring can be employed to indicate possible impending failure well before it becomes a significant probability. The advantages to be obtained from the use of condition monitoring are shown in *Table 9.1.*[1]

There are two major reasons for not employing condition-based maintenance in certain circumstances. First, not all causes of plant failure can be detected in advance. If the more probable sources of failure of a unit fall into this category condition monitoring will be of little value. Second, condition monitoring is, by its very nature, costly in manpower, monitoring equipment or both. Only if the cost of

* Contributed by T. A. Henry, Senior Lecturer in Engineering, University of Manchester.

148

Table 9.1. ADVANTAGES PROVIDED BY CONDITION MONITORING

	Advantages obtained	Methods by which condition monitoring gives these advantages	
		Trend monitoring	Condition checking
Safety	Reduced injuries and fatal accidents to personnel caused by machinery	Enables plant to be stopped safely when instant shutdown is not permissible	Machine condition, as indicated by an alarm is adequate if instant shutdown is permitted
	More running time	Enables machine shutdown for maintenance to be related to required production or service, and various consequential losses from unexpected shut downs to be avoided	Allows time between planned machine overhauls to be maximised and, if necessary, allows a machine to be nursed through to the next planned overhaul
Increased machine availability	Less maintenance time	Enables machine to be shut down without destruction or major damage requiring a long repair time. Enables the maintenance team to be ready, with spare parts, to start work as soon as machine is shut down	Reduces inspection time after shutdown and speeds up the start of correct remedial action
Output	Increased rate of net output		Allows some types of machine to be run at increased load and/or speed. Can detect reductions in machine efficiency or increased energy consumption
	Improved quality of product or service	Allows advanced planning to reduce the effect of impending breakdowns on the customer for the product or service, and thereby enhances company reputation	Can be used to reduce the amount of produce or service produced at sub-standard quality levels

Table 9.2. SUMMARY OF CONDITION MONITORING TECHNIQUES

Method	On/off load	Location of fault	Equipment costs (1977 level)	Skill required of operator	Comments
1. Visual	On	Surface only	nil	Predominantly experience	Covers a wide range of *ad hoc* methods
	Off	Can be extended to interior components provided method was considered at design stage	Optical probes about £500 £2000 television	No special skill required	Extensively used in aero engine industry for 'turn round' inspections
2. Temperature (General purpose technique)	On	Surface or internal	Varies widely	Little skill required for most methods	Instruments range from direct reading thermometers to infra-red scanners
3. Lubricant monitoring (General purpose technique)	On	Any lubricated component – via magnetic plug, filters or oil samples	<£50 except for ferrography and spectrography equipment	Skill is required to distinguish between damage debris and normal wear debris	Spectrographic and ferrographic analysis services are available to show what elements are present
4. Leak detection	On and Off	Any pressure-containing component	<£1000	Skill in use of the specialised equipment readily acquired	
5. Crack detection (a) dye penetrant	On and Off	At clean surface	<£50	Some skill required	Only detects cracks breaking surface
(b) magnetic flux	On and Off	Near to clean smooth surface	<£50	Some skill required. Easy to miss crack	Limited to magnetic materials. Sensitive to crack orientation
(c) electrical resistance	On and Off	At clean smooth surface	<£100	Some skill required	Sensitive to crack orientation. Useful for estimating crack depth
(d) Eddy current	On and Off	Near to surface. Closeness of probe to surface affects results	£100–£1000	Skill essential	Detects a wide range of material discontinuities, cracks, inclusions, hardness, etc.

Technique	On/Off	Access/Application	Cost	Skill	Remarks
		...ponent to which there is access via a clean smooth surface		cracks not to be overlooked	...fore general searches lengthy. Used to back up other diagnostic techniques
(f) Radiography	Off	Access to both sides necessary	>£1000 (Battery operated)	Considerable skill required in setting up and interpreting radiographs	Covers a large area at one time. Security required because of radiation hazard. Limited to sections less than 50 mm (steel)
6. Vibration monitoring (general purpose technique), total signal, band frequency analysis, or peak level	On and Off	Any moving component. Any object containing moving parts. Transducer placed in path of vibration transmission, e.g. bearing housing	>£500	Some skill required	Methods vary from the simple to the sophisticated. Routine measurements taken rapidly and do not affect operation of the machine
7. (a) Corrosion monitoring		In pipes and vessels			
(b) Corrosometer (electrical element)	On		Potentiometer <£200	Some skill required	Will detect 1 μm corrosion loss
(c) Polarisation resistance and corrosion potential	On		Meters £500	Some skill required	Only indicates that corrosion is occurring
(d) Hydrogen probe	On		£100	No skill	Hydrogen evolved diffuses into thin walled probe tube and causes pressure rise
(e) Probe indicator holes	On		–	Skill required in drilling to exact depth	Indicates when preset amount of corrosion has occurred
(f) Weight loss coupons	Off		–	–	Monitored when plant stripped down.
(g) Ultrasonics	Off		£500–£1000	Skill essential	Will detect 0.5 mm thinning

condition monitoring is lower than the expected reduction in maintenance labour costs and unavailability costs, or where safety of personnel is a relevant consideration, is it worth applying condition based maintenance. However, it is worth noting that in the process industry it is not uncommon for the annual maintenance cost to exceed the purchase price of the plant, so there is plenty of incentive for reducing maintenance costs.

9.1 Categories of Condition Monitoring

Condition monitoring falls into two distinct classes: (a) monitoring which can be carried out without interruption to the operation of the unit, and (b) monitoring which requires the shutdown of the unit, or at least, the release of the unit from its prime duty. Clearly, the former category has significant advantages over the latter since no interruption to plant output is involved. However, there are many situations where the monitored unit is shut down regularly or frequently as part of the normal plant operating policy, for example two-shift power stations. In such circumstances off-load monitoring need not interrupt normal plant operation.

9.2 Methods of Condition Monitoring

Most techniques of condition monitoring amount to the systematic application of commonly accepted methods of fault diagnosis. The range of methods in common use is very wide. Certain methods tend to be associated with particular types of plant or particular industries. In the following sections only the more common methods will be mentioned. A summary of condition monitoring techniques is shown in *Table 9.2.*

9.2.1 On-load Monitoring Techniques

Visual, aural and tactile inspection of accessible components Looseness in non-rotating components can be detected readily. Wear debris or fretting corrosion debris from friction joints such as bolted, riveted or shrink-fit joints, is a clear indicator of looseness. Relative motion as small as 1 μm at the interface between two components can be sensed by finger touch. Brittle lacquers can be applied to joints to provide indication of relative motion. Loose moving components can often be detected audibly and loose joints respond to tapping with a dull heavily damped sound. Inaccessible internal parts of machines can be examined using borescopes or similar optical aids[2].

Temperature monitoring The temperature sensitive aspects of a unit can readily be monitored. Temperature sensors include contact thermometers, thermocouples, thermistors, temperature chalks and paints, and infra-red detectors. Two examples where temperature monitoring can give warning of mechanical trouble are lubricant temperature at bearing outlets and engine cooling water temperature. This is a general purpose technique and is described further in Section 9.6.

Lubricant monitoring The use of magnetic drain plugs in unit lubrication sumps is well known. The existence of magnetic debris gives information on the surface condition of load bearing parts. Examination of the oil and filters will reveal metallic debris in suspension or deposited on the filters. Both the wearing-in of a new gear train and the onset of contact stress fatigue are accompanied by the removal of metal but the shape of the metal removed is very different in the two cases. This is a 'general purpose' technique of monitoring which is being developed in a sophisticated manner[2, 3, 4, 5]. Lubricant monitoring is described further in Section 9.5.

Leak detection A number of leak detecting techniques are available, including the soap and water methods. The use of proprietary preparations can make this method more effective, capable of detecting leaks as low as 1 μl/s. One powerful technique is ultrasonic detection[6]. When a fluid is forced through a leak under internal or external pressure, sound is generated in the frequency range 40–80 kHz. The ultrasonic leak detector identifies this very high frequency which is easily separated from the lower frequencies of ordinary machine noise. A typical detector can sense the presence of a 50 μm dia. hole in a unit containing a pressure of 0.1 bar, at, say, 10 m distance, or expressed as a flow rate, 10 μl/s. Halogen testing, where a search gas, e.g. Arcton 12, is inserted into the system under test and the presence of the search gas outside the system indicates a leak, can be used to detect leakage rates as low as 1 pl/s. Unfortunately, it is common to find sufficient halogens free around a plant to mask the effect of the leak.

Vibrating monitoring   On-load vibration monitoring can be used to detect a wide range of faults in machinery. This method has wider application than any other monitoring technique. For example, vibration measurements made near the bearings of a machine can detect and distinguish between imbalance, shaft misalignment, damaged bearings, damaged gears and other transmission components, mechanical looseness, cavitation or stall and a number of other faults. Although the basic methods of monitoring are straightforward, in many cases very much more information can be extracted from the measurements if modern signal processing techniques are applied. This is a field which is currently receiving much attention and consequently understanding and practical experience are improving rapidly. Vibration monitoring is examined in detail in Section 9.7.

Noise monitoring   Apart from the detection of special sounds as in leak detection, noise monitoring can be applied in the same manner as vibration monitoring. However, since a noise which is an indicator of the condition of a unit must have originated as a vibration of some part of the unit, it is usually more effective to monitor the original vibration. There are, however, situations where noise monitoring is more convenient. The methods are discussed in more detail in Section 9.7.

Corrosion monitoring   Special electrical elements placed in a system change resistance as corrosion progresses. Using special probes, corrosion rate can be indicated from the polarisation resistance of the probe, whilst simple measurement of the potential between a reference electrode and the system will indicate if corrosion is occurring. Evolution of hydrogen during corrosion can be detected within a thin walled blind tube into which the hydrogen diffuses. Drilling of fine bore holes from the outside nearly through the vessel wall will give warning of pre-determined amounts of corrosion.

9.2.2   Off-load Monitoring Techniques

Visual, aural and tactile inspection of normally inaccessible or moving parts   The condition of many transmission components can readily be checked visually. For example the wear on the surfaces of gear teeth gives much information. Problems of over-load, fatigue failure, wear and poor lubrication can be differentiated from the appearance of the teeth. A wide selection of devices is available to assist inspection, including borescopes, fibre optic introscopes, tank periscopes, mirror sets and closed circuit television cameras.

Crack detection Most serious failures are preceded by crack growth from a point of stress concentration or from a material defect at the surface of the component. It is widely believed that fatigue failures occur without any warnings. This is untrue, though the warning crack is not normally visible on casual inspection. To overcome this difficulty a number of crack detection techniques have been developed[2,6].

1. *Dye penetration into surface cracks.* Causes cracks as small as 0.025 μm to be revealed to the naked eye[6].

2. *Flux testing of magnetic materials.* A crack or other defect, which crosses the path of a magnetic field (which is induced locally into the surface of the material using U-type magnets), causes the magnetic flux to spread round the crack into the space above the surface. The existence of this field and hence of the crack is revealed using magnetic powder[6].

3. *Electrical resistance testing.* The presence of a crack at the surface will increase the resistive path measured between two probes in contact with the surface. In spite of difficulties with surface contact this method can be used to detect and measure the depth of cracks.

4. *Eddy current testing.* A coil carrying current placed close to the surface induces eddy currents in the material. These eddy currents are detected either by a change in the inductance of the generator coil or by a separate search coil. Although it is not necessary to have a clean smooth surface problems arise in interpreting results.

5. *Ultrasonic testing.* Ultra sound generated at the surface of the component will be reflected at any surfaces in the path of the sound whether they be manufactured or faults. The time delay between generation of the sound pulse and detection of the reflection is displayed to give a measure of the distance of the surface from the source. The generation and propagation of ultra sound in the frequency range 0.25–10 MHz is very directional so suitable orientation of the transmitter enables imperfections such as cracks to be distinguished from the outside surfaces of the component. The technique has been developed to a high degree of sophistication. Cracks can be detected in a weld, and cracks in the surface of a shaft can be detected by examination from one end or from a suitable step. The range of operation is 5–15 mm, in steel[2].

6. *Radiographic examination.* Imperfections can be photographed using either X-rays or gamma rays from a radioactive source and special photographic materials. A change in thickness of 2% can be detected.

Member thickness is normally limited to 50 mm. The method usually requires dismantling of the unit being examined and raises problems associated with the protection of personnel.

Leak detection Ultrasonic leak detection can be applied to units off load by placing an ultrasonic generator inside the component to be tested.

Vibration testing The response of a unit to vibration forcing can reveal much information. One of the most common vibration tests for rotating machines is the run-down test which takes advantage of system resonances to magnify vibrations.

Corrosion monitoring In addition to the methods described under on-load monitoring, corrosion rate can be determined more accurately by placing 'coupons' in the plant and removing them for weighing at intervals. Ultrasonic measurements will detect the change in dimensions due to corrosion. A new technique for remote measurement of corrosion thickness employs a pulsed laser to 'drill' a fixed depth per pulse and a light detector to indicate when the hole has reached the bright metal.

9.3 General Purpose Monitoring Techniques

Only three of the condition monitoring techniques described in Section 9.2 and shown in *Table 9.2* can be considered as true 'general purpose' monitoring methods. These are thermal, lubricant and vibration monitoring. In each of the three the parameter being monitored contains information that has been transmitted through the machine. Hence a change in a parameter detected at an appropriate monitoring point can indicate a change in any of a number of different components. In contrast other monitoring methods, e.g. crack detection, require monitoring of the actual component that is at risk. Hence such monitoring methods would be too time consuming and expensive to be used except to examine particular parts of a component which is known to be particularly prone to fail. *Table 9.3* provides a comparison of the three general purpose monitoring methods.

Vibration monitoring, which is the most versatile technique for condition monitoring, is described in detail in Section 9.7. So as to provide a complete picture lubricant and thermal monitoring are described briefly in Sections 9.5 and 9.6.

Table 9.3. COMPARISON OF GENERAL PURPOSE CONDITION-MONITORING TECHNIQUES

	Thermal monitoring	Lubricant monitoring	Vibration monitoring
Medium for transmission of information through machine	Solid – casing, shaft body Fluid – lubricant, cooling water or air Depends on thermal conductivity	Oil used for lubrication and/or cooling Depends on lubricant being pumped round the machine	Any solid part of machine Depends on elastic and mass characteristics of solids
Components monitored	Any heat generating devices (combustion in cylinder or electrically generated heat in motor). Condition of bearings. Fluid flow in heat exchangers (fouling of passages)	Any component which is lubricated; bearings, transmission components (gears, couplings, cams), lubrication pump	Any component that moves, surfaces between components with relative motion, clearances
Faults detected	Failure of drives, blockage of ducts, loss of cooling, fouling of coolers, over-use (e.g. overloading motors)	Any form of wear or failure that results in lubricated surface failure. Leakage of other contaminants into lubricant	Change in any moving components, wear or failure of bearings, mis-balance, change in clearances
Monitoring equipment	Fluid or bimetallic thermometers, resistance thermometer, thermistor plus associated instruments, temperature paints/crayons, infra-red detectors, optical pyrometers, infra-red scanning camera	On-load removable filters, magnetic plugs for visual examination of debris using microscope, spectroscope for analysis of material in suspension, ferroscope for separating debris, pressure gauge across filters	Accelerometer plus electronic processing equipment to display time averaged values. Frequency filters and recorders for analysis of vibrations
Frequency	Continuous and periodic	Primarily periodic	Periodic but also continuous

9.4 The Systematic Application of Condition Monitoring

One of the problems of condition monitoring is the state of confusion generated by the wide variety of monitoring techniques. Even within the field of vibration monitoring, as we shall see later, there is a profusion of measurement methods, each with its particular virtue. Each technique requires a special skill and understanding, and some of the more useful techniques have been developed only in the last decade.

Two particular problems face the engineer embarking on a programme of condition-based maintenance. How rapidly does a plant malfunction develop and which technique, if any, will best detect the types of component failure experienced in the plant? Trend monitoring assists with the first problem whereas a formalised approach to monitoring assessment, such as the LEO method, can help considerably with the second.

9.4.1 Keeping Records

The key feature of condition monitoring, whatever the technique in use, is the accumulation of data, which at the time of collection seem to have limited value. However, only when sufficient data are collected can normal condition and possible deviations be established and redundant data cease to be accumulated. Thus keeping records is essential to any monitoring programme. To keep paper work to a minimum the data recording system should be devised to provide the important information clearly to the decision maker. Where possible the forms devised for collection of data should also serve to present the information necessary for decision making so that transfer of data is avoided. For example, where the trends in a monitored parameter are used to indicate plant condition it is useful for the data to be plotted directly onto graphs since the mind can absorb visually presented data more effectively than written data. Unfortunately, in the early stages of a monitoring programme there is considerable redundancy in measurements. Only when experience of normal conditions and malfunctions has been obtained will the engineer be able to cut down on the number of measurements made. The importance of recording all relevant particulars of the machine and of the monitoring equipment and tests cannot be over emphasised. One of the most common difficulties encountered by the decision maker, when a monitoring technique detects a trend indicative of failure, is lack of data. Where many machines are operated consideration should be given to the advantages of computer data storage and presentation.

9.4.2 Trend Monitoring – Life Curves

If sufficient data of monitored parameters are available for a group of similar machines it is possible to relate the condition of the machines directly to the monitored parameter. Unfortunately this procedure has shown limited success, because parameter levels vary widely between nominally identical machines. Acceptable and unavoidable tolerances in components, dimensions, surface finishes, etc., will cause significant variations in many monitored parameters. A more successful technique is to monitor trends in parameter levels.

The life of a machine can be considered to be controlled by two distinct groups of defects. Early life defects, such as inadequate clearances, roughness in mating surfaces, debris left from machining or assembly, tend to clear themselves creating what is known as the 'running-in-period'. During this period the monitored variable, for instance temperature or vibration level at a bearing, or rate of accumulation of debris at a magnetic plug or at a filter, will decrease in level with time. During this early life period the machine should be carefully monitored for a change from decay to growth since probability of failure is high initially, decaying as the defects are cleared during running-in. Final life defects,

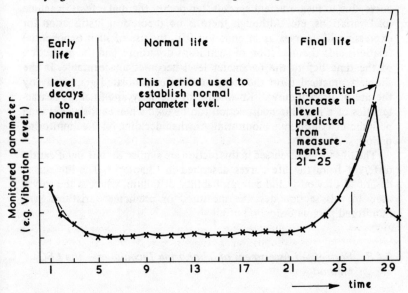

Figure 9.1. Typical monitored parameter – life curve condition. Monitoring engineer recommends repair at measurement 26. Repair carried out between measurements 28 and 29.

such as overload, looseness, wear or metal deterioration, result in an increasing probability of failure with time accompanied by an increase in the monitored parameter.

The monitored parameter will usually vary in a consistent manner, decaying in early life, remaining constant or rising slowly in mid or normal life and increasing in final life (*Figure 9.1*). In order to assess abnormal trends, the decay in early life and the increase in final life can be considered to be exponential. Collacott[13] has considered reciprocal and exponential relations in the probability of failure of machines and Sanker and Xistrus[14] found exponential trends in vibration monitoring data obtained by the Canadian Navy. Thus the monitored parameter of a machine through its life could be represented by the expression

$$V = E \exp\left(-t/e\right) + M + F \exp\left((t-T)/f\right) \qquad (9.1)$$

where E and e control the vibration decay in early life, M the vibration in mid-life and F and f the growth in vibration during final life leading up to machine failure near time T. This expression describes the typical 'bath tub' life curves characteristic of many monitored variables (*Figure 9.1*). Actual measurements will show a scatter about such a curve, due to measurement spreads, temporary thermal effects, machine load variations, etc. Although there is no theoretical justification for expressing life curves as in equation 9.1, the use of such exponential relations aids the detection of significant deviations and the prediction of the time before the parameter level becomes unacceptable. In the simplest practical form the trends can be extrapolated graphically by the use of French curves. Knowledge of the early and final life characteristics of a machine malfunction can be used when choosing between periodic or continuous monitoring or when deciding on the monitoring period.

The life curves discussed in this section are similar to, but significantly different from the life curves described in Chapter 3. The life curves described previously indicate probability of failure, whereas the curves used in this section describe measured or predicted variations in a monitored parameter.

9.4.3 Formalised Assessment of Monitoring Techniques — the LEO Method

The choice of the monitoring technique to be applied in a particular plant is difficult, affected by such factors as skills available, effect of the environment, reliability of the machines and the monitoring methods, etc. At present this decision tends to be made on an *ad hoc* basis of personal experience or advice. There is, however, a clear requirement for a formalised approach to be developed.

In the field of 'non-destructive testing' where a variety of competitive techniques for surface crack detection have long been established, the LEO approach has been put forward (Birchon[2, 7, 8]) as a simple formalised way of selecting the appropriate technique by assessing effectiveness. Three separate factors are considered and then combined to provide a measure of the size of fault that must be detected for the technique to be effective.

(1) The critical parameter size, L, which will cause the particular failure being considered. In the context of cracks this could be the crack depth determined by calculation, experiment or studies of failures. If the failure were seizure of a rolling bearing it is unlikely that the size of the damage in the races would be considered. More appropriately it would be a correlation between vibration level and occurrence of seized bearings. Thus, L would be the vibration level at which experience has shown the bearing will seize.

(2) The experience factor, E, to take account of:

(a) the amount of knowledge possessed of the cause and mechanism of the failure and,

(b) the rate of progression of the failure in relation to monitoring frequency.

Both these points of experience amount to a measure of the extent of previous trouble encountered.

(3) The monitoring operational efficiency factor, O, which takes account of:

(a) the sensitivity and reliability of the method and the equipment in relation to the characteristics of the machine and its environment, and

(b) the skill of the monitoring personnel.

O would vary from one, where failures had been thoroughly monitored, to zero where it had never been possible to detect the fault in the past.

The product, $L \times E \times O$, provides a measure of the monitored parameter level or change which the monitoring technique must be able to detect to be certain of providing sufficient warning of an impending failure.

Unfortunately, although data have been collected for NDT applications[2, 7, 8], the approach has not yet been taken up for general condition monitoring and no information is available on L, E or O

factors applicable to thermal, lubricant or vibration monitoring. However, the maintenance engineer will find it useful to develop his own factors for the techniques he has at hand. Some proposals for monitoring operational efficiency factors, derived from Birchon[2] are shown in *Table 9.4.*

Table 9.4. PROPOSED MONITORING OPERATIONAL EFFICIENCY FACTORS

		Subtract from $O = 1.0$
Experience of monitoring personnel	None	0.2
in applying the technique	Some	0.1
	Considerable	0
Conditions of access in relation to	Poor	0.2
the requirements of the technique	Fair	0.1
	Good	0
Conditions of the environment	Bad	0.2
affecting the technique (noise,	Average	0.1
dirt, fumes, heat, electrical inter-	Good	0
ference, use of protective clothing)		
Availability of direct comparisons	None	0.2
(standard lubricant debris, vibration	Some	0.1
spectra, temperature graphs, etc.)	Sufficient	0

9.4.4 A Simple Example of the LEO Approach

Cracks are known to develop in a rotor shaft at a shoulder located approximately at mid span. Some previous experience of vibrations caused by such crack development has been obtained and shafts containing cracks have been removed from service and examined. No measurable indications have been found at running speed but on run-down through the critical speed the resonant vibrations are affected by the crack which changes the stiffness of the shaft in one plane. Fracture mechanics analysis shows that the crack affects the critical speed as shown in *Figure 9.2,* and it is first decided to determine whether a change in critical speed could be used to detect the crack sufficiently early. It is decided that the monitoring technique must detect the crack growth before the crack depth has reached 0.5 of the shaft diameter, D. Experience has also shown that once the crack depth reaches $0.1D$ it develops at $0.1D$ per week of running. The run-down tests require that the machine be taken off load, which can prove costly, so the minimum acceptable period between tests is fixed at one week.

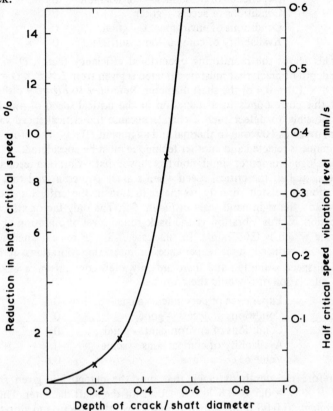

Figure 9.2. Effect of a fatigue crack on the vibration characteristics of a rotor

Since the crack depth must not exceed $0.5D$, let $L = 0.5$. Some experience of the cracks exists and the progression rate is known fairly precisely. Hence it would be sufficient to set E to take account of the maximum crack growth possible between tests, i.e.

$$E = \frac{0.5 - 0.1t}{0.5}$$

where t is the number of weeks between tests. For weekly tests $E = 0.8$.

The method of monitoring is routine, using vibration transducers mounted on the bearing pedestals, so according to *Table 9.4* deductions from the monitoring operational efficiency factor would be as follows:

Experience of personnel — considerable	0
Conditions of access — good	0
Conditions of environment — good	0
Availability of comparison — sufficient	0

This leaves the monitoring operational efficiency factor, $O = 1.0$. The depth of crack that must be detected is given from $L \times E \times O = 0.5 \times 0.8 \times 1.0 = 0.4$ of the shaft diameter. Reference to *Figure 9.2* shows that this corresponds to a reduction in the critical speed of 4%. It is not possible to detect such a change because the critical speed varies by as much as 10% due to thermal and alignment effects. Therefore this technique is rejected and another technique must be considered.

Analogue computer simulation has shown that a vibration resonance appears at half the critical speed when a crack is present and that the vibration is related directly to the reduction in the critical speed as shown by the right-hand scale in *Figure 9.2*. The only factor affecting detection of this vibration is the background level of vibration at the bearing which is 0.04 mm/s. In this case L and E remain unchanged but O is reduced since experience in measuring vibrations at half critical speed is limited and there are only a few comparisons available. The calculation of O would therefore be:

Experience of personnel — some	−0.1
Conditions of access — good	0
Conditions of environment — good	0
Availability of comparisons — some	−0.1
Value of O	0.8

Therefore the depth of crack that must be detected is given from $L \times E \times O = 0.5 \times 0.8 \times 0.8 = 0.32$ of the shaft diameter. This is equivalent to 0.08 mm/s which is well above the background vibration level, hence weekly monitoring of the half-critical speed resonance would be satisfactory.

Less frequent tests would be more economical but frequency of tests affects the experience factor, E. If the smallest detectable increase in vibration at half-critical speed is equal to the background level, i.e. 0.04 mm/s equivalent to a crack depth of $0.23 D$, then

$$L \times E \times O = 0.5 \times E \times 0.8 \geqslant 0.23$$

$$E = (0.5 - 0.1t)/0.5 \geqslant 0.575$$

and

$$t \leqslant 2.125 \text{ weeks}$$

Fortnightly run-down tests would therefore detect a crack sufficiently soon using this method.

From this example it can be seen that the LEO approach formalises the method of assessing a technique by splitting, for separate consideration, the various relevant factors. When this or a similar approach becomes established in condition monitoring the added advantage of being able to obtain *L, E* and *O* factors from tables will greatly enhance its value.

9.5 Lubricant Monitoring

It is not possible to examine the working parts of a complex machine on-load, nor is it convenient to strip down the machine. However, the oil which circulates through the machine carries with it evidence of the condition of parts encountered. Examination of the oil and any particles it has carried with it allows monitoring of the machine on-load or at shut-down. A number of techniques are applied, some very simple, others involving painstaking tests and expensive equipment.

9.5.1 Lubricant Monitoring Techniques

Lubricant examination[4] can cover the debris deposited, the debris in suspension, or the condition of the oil.

Debris deposited   The larger particles carried along by the lubricant can be collected on-load in filters or magnetic collectors.

1. *Filters.* The rate of build-up of debris on a filter is readily monitored on-load by measuring the pressure drop across the filter. Removal of the filter, which can be carried out on-load if the machine is suitably designed, and subsequent examination of the debris under the microscope to establish size and shape or with a spectrograph to determine the element content, provides a more sophisticated method of detecting a significant change in any component 'visited' by the lubricant. It is good practice to store the debris on an adhesive substrate as part of the routine condition records.

2. *Magnetic debris collectors* are a convenient way of capturing ferrous components. Magnetic plugs can be designed to be readily removed on-load and one proprietary magnetic debris collector can be monitored without removal, giving an indication of debris build-up.

Debris in suspension   The smaller particles collected by the lubricant will remain in suspension. It is claimed that examination of the particles in suspension gives the earliest warning of component damage. Quantitative measurements can be made but account has to be taken of loss of lubricant during running and the dilutive effect of topping-up with debris free fresh oil.

1. *Spectrometric oil analysis* (SOA). The concentration of critical wear materials in the oil is determined by an emission spectrometer or an atomic absorption spectrometer both of which measure concentration of elements. This process is rapid and convenient provided there is ready access to a spectrometer. However, it provides only information on rate of build-up and composition of debris, and no information on shape of debris.

2. *Ferrographic oil analysis*[5] is a means for depositing magnetic particles distributed according to size from an oil sample onto a substrate. The particles can then be examined for concentration, size distribution and shape.

Condition of used oil   The oil itself can be examined more generally for indication of other malfunctions. Some of the indications and their causes are listed below:

Indication	Cause	Action
Foaming	Excess churning or passage under pressure through a restriction	Check system
	Detergent contamination	Change oil
Emulsion separates out naturally	Water ingested	Drain off water
separates with centrifuge	Water	Change oil
Colour darkened	Oxidation of oil, excess temperature	Change oil
	Combustion or other products reaching oil	

9.5.2   Malfunctions that can be Detected by Lubricant Monitoring

A change in the rate of debris collection indicates a change in the condition of the machine. During running-in the debris collection rate will reduce with time unless a failure occurs. During normal life the rate of debris collection and the composition, size and shape of the debris will remain constant. When a change occurs, knowledge of the change in composition will assist in determining which component has changed.

Normal wear particles tend to be flat, whereas cutting or abrasive wear results in spiral shaped debris. Surface fatigue failures produce larger angular particles.

Simple analysis of rate of debris collection in filters or magnetic plugs will indicate bearing damage and damage to any load bearing or sliding surface, such as gear faces or other transmission components. Use of spectrometry or ferrography will enhance this monitoring with data which will assist in pinpointing the damaged component.

9.6 Thermal Monitoring

Monitoring the temperature of a component in a machine is undertaken for one of three purposes.

1. To enable the temperature of a process to be controlled manually or to check that it is being controlled properly.

2. To detect an increase in heat generation due to some malfunction such as a damaged bearing.

3. To detect a change in the heat transmitted through and out of the body of the machine caused by a change in some component, such as a failed coolant circulator or ash build up in a boiler.

The first listed purpose is widely applied but the use of thermal monitoring for general malfunction detection is not so widely used.

9.6.1 Location of Temperature Measurements

Monitoring can be carried out at a point within the body of the plant, e.g. water temperature in a boiler, or at the surface of a component, e.g. a bearing housing. Surface measurements provide general information on heat generation within a machine and on heat transmission paths to the surface or to heat exchangers. Unfortunately measurement of surface temperature is more difficult than of immersion temperature since the sharp discontinuity of the temperature profile which usually occurs at the surface is easily modified by the installation of a temperature sensor. Sensors for surface measurement are therefore restricted to small devices such as thermocouples or non-contact methods such as radiation meters.

9.6.2 Temperature Monitoring Devices

Contact sensors Devices which take up the temperature of the body with which they are in contact, and then transmit information on their own temperature, are most commonly used. They can provide temperature indication locally or remotely (thermometers) or control some function from the temperature (thermostats). Both accuracy and response times of contact sensors are affected by attachment methods. Good thermal contact is important. This is best provided for surface measurement by embedding or welding the sensor to the body. Response times are related to the volume of the sensor so the smaller devices are more suitable if temperatures vary rapidly.

1. *Liquid expansion sensors* are the most commonly used devices. Mercury or alcohol in glass thermometers are accurate but fragile devices. Liquid in metal sensors are used where indication is required some 0.5–2 m from the sensor. All these sensors are large (up to 6 X 12 mm) and are therefore unsuitable for surface measurement.

2. *Bimetallic expansion sensors* can be made compact and are widely employed in thermometers where temperatures are high or where a robust device is needed. These devices are unsuitable for surface measurement and generally less sensitive than liquid expansion devices.

3. *Thermocouple sensors* are the smallest and most adaptable temperature devices. When two wires of different metals are joined a thermoelectric potential is generated across the junction, proportional to temperature. This potential can be measured by a simple voltmeter, a potentiometer or an electronic voltmeter. Since the circuit must be completed through the meter it will contain two opposing junctions between the dissimilar metals. Thus the potential at the meter will indicate the temperature difference between the two junctions. This reference junction could be at the terminals of the meter or mounted in a reference body. The sensitivity of the commonly used thermocouple pairs, copper/constantan or chromel/alumel, is about $40 \ \mu V/^{\circ}C$ difference in temperature and these devices can measure to about $0.5^{\circ}C$ accuracy using a potentiometer or an electronic voltmeter. A number of compact battery-portable and mains-powered meters for thermocouple measurements are available.

The junction can be as small as 0.5 mm across so, provided appropriate precautions are taken to minimise wire conduction error, thermocouples can be used in regions where the temperature gradient is very steep. They are suitable for surface measurements if correctly attached. They can be used at high temperatures (Cu/Con $300^{\circ}C$, Chromel/Alumel $700^{\circ}C$) but are better protected from corrosion if sheathed. It is common to use copper sheathed wires for protection from damage. Copper sheathed mineral insulated thermocouples are available in diameters down to 2.5 mm.

4. *Resistance sensors* make use of the change of resistance of an element with temperature. The element may be a wire or a thin film glued to, or deposited on, the surface. Since the resistance of most common metals changes by only 0.003/°C these devices are very insensitive and require a bridge circuit and sensitive meter. Thermistors are semiconductor devices whose resistance changes rapidly with temperature. A typical thermistor bead encapsulated in glass measures 1.5 mm in diameter and is about ten times more sensitive than metal sensors. The temperature range is limited to about 300°C and the resistance tends to drift with age requiring rezeroing. In a typical bridge using a 1 V supply a metal sensor will produce an output of about 1 mV/°C and a thermistor about 10 mV/°C. A typical application of thermistor sensors is the measurement of cooling water temperature in i.c. engines. Compact thermometer units based on thermistor sensors are available for use as portable instruments.

Temperature paints, crayons, and pellets A simple method of monitoring surface temperatures is the use of paints, crayons or papers which change colour at known temperatures. Indicators are available with reversible colour changes and with irreversible changes for indicating the maximum temperature reached. Pellets indicate temperature by melting. Sets of materials are available giving a range of indicating temperatures of 40–1400°C in steps varying from 3°C at the low end to 30°C at the high end of the range. Accuracy is about 2% or 5°C.

Non-contact sensors Energy radiation from a body varies with the absolute temperature of the body, T, and the emissivity of the radiating surface, e, according to the Stefan-Boltzmann law

$$E = \sigma e T^4$$

This enables surface temperature to be deduced, from the energy radiated, without any direct contact. The major cause of inaccuracy is the variation in the emissivity. The range of measuring devices, described in order of ascending cost, is as follows:

1. *Optical pyrometer.* At about 500°C a significant part of the radiation is in the visible frequencies. This phenomenon is used in the optical pyrometer where the colour of the body surface, or of the gas,

is compared with a heated filament to give a reading of temperature to within 2%.

2. *Radiation pyrometer*. This uses thermopiles or lead-sulphide cells to measure the radiant-energy received from a hot surface, either in a particular frequency band, such as infra-red, or over the whole spectrum. The temperature is then indicated on a meter to an accuracy of about 2% over the temperature range 50–4000°C. The viewing angle of these devices varies from 3° to 15°.

3. *The scanning infra-red camera*. This scans the field of view and displays the thermal profile as a grey scale on a TV monitor. These units cover a temperature range of 20–2000°C and can give a resolution as low as 0.2°C at 20°C. Such devices are very expensive but have valuable applications in the search for hot spots.

9.6.3 Malfunctions that can be Monitored Thermally

In addition to the prime function of monitoring a process or system temperature as a check that the controls are working correctly, there is a range of general faults that can be detected by thermal monitoring.

1. *Bearing damage*. Damage to the elements in a rolling bearing, or to the surface in rubbing or hydrodynamic bearings, will result in an increase in heat generation. Provided that the bearing does not incorporate a thermostatically controlled cooling system, this increase in heat generation will result in a rise in the temperature at the surface of the bearing housing. This can be detected by a surface mounted sensor such as a thermocouple or better still from the difference in temperature indicated by a pair of sensors, one mounted at the surface and the other a small distance below the surface.

Any surface contact which has come about as the result of damage or wear will generate heat which must be transferred to some external surface in order to escape and this can be detected at that surface.

2. *Failure of a coolant*. Failure of the lubrication or coolant will be detected by a temperature rise at the appropriate body surface. Such failures can result from a pump failure due to an internal fault or a drive fault, blockage of a pipe, valve or filter, or a damaged heat exchanger.

3. *Incorrect heat generation.* Incorrect combustion in an i.c. engine or a fossil fuel boiler can cause an uneven temperature distribution at the casing surfaces. A series of suitably located thermocouple sensors whose outputs are scanned and recorded will show up unevenness or

changes in the distribution. Temperature paints or a scanning infra-red camera can be used to monitor large areas rapidly.

4. *Build-up of unwanted materials.* The build-up of sludge or sediment in pipes, of ash or dust in boilers or ducts, and of corrosion by-products, will all be detected by temperature scanning of the appropriate surfaces since the build-up increases the insulation of the component and alters the surface temperature.

5. *Damage to insulating materials.* Where an item of plant incorporates insulation, damage to the insulation is readily detected using the scanning infra-red camera. Cracks in refractory linings and damage to lagging will show up as hot or cold spots.

6. *Faults in electrical components.* Where a current-carrying connection is poor the contact resistance results in extra heat generation which is readily detected with a scanning infra-red camera. For example, high voltage transmission lines are monitored regularly for faults in cables, connections and insulators using a scanning camera operated from a helicopter. Failed components such as rectifiers, thyristors or windings can be detected as cold spots.

9.7 Vibration and Noise Monitoring

This is perhaps the oldest form of machine condition monitoring. Everyone has at one time or another noticed a change in the noise produced by the family car or mowing machine, or has been disturbed by the vibrations of the central heating system. Such changes in the heard or felt vibrations automatically spark off a period of close and worried examination of the offending machine. In the ensuing sections modern vibration and noise monitoring methods will be described together with the fundamental concepts of vibrations and their analysis essential for the understanding of these techniques.

Changes in noise or vibration indicative of trouble are not necessarily increases in level. A failed belt drive to a cooling fan could result in an immediate reduction in fan noise long before overheating causes an increase in vibration together with serious damage. Thus in the application of vibration or noise monitoring as a preventive maintenance tool it is changes that are of particular interest.

9.7.1 The Cause of Vibration and Noise

All machines vibrate. It is difficult to balance moving parts, so vibrations originate in unbalanced rotating parts and in accelerations of components

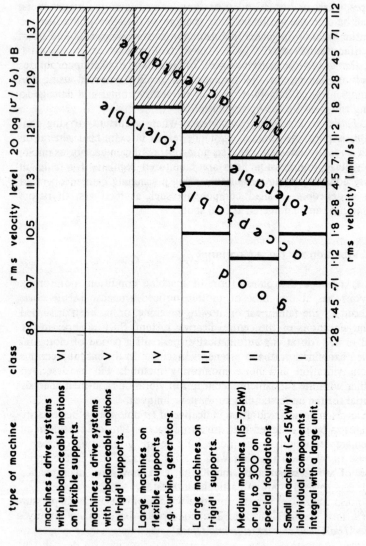

Figure 9.3. Vibration severity related to classification of machine. ($v_0 = 10^{-5}$ mm/s) – BS 4675, Part 1, 1976. This chart should not be used as a criterion for vibration condition monitoring

moving along linear paths. Components which move, rub, or roll on adjacent components generate vibrations because of the roughness in the mating surfaces. Looseness in the fit between mating components results in impacts. Load bearing components subjected to load cycles deflect under the loads and hence transmit vibrations. Standards of vibration based on generally accepted qualities of workmanship are laid down in BS 4675 : 1976. These standards, which are shown in *Figure 9.3*, may be used to assess the standard of workmanship and design of a machine or to assess its general condition. However, they cannot be applied as criteria in relation to plant condition monitoring since the criterion bands are wide in order to allow for production spreads in tolerances and the effects of differing foundations.

Machine vibrations cause noise, the level of which depends on the surface area of the vibrating parts and the efficiency of transmission of noise from the machine. For health and environmental reasons much effort is being devoted to reducing the vibrations in machines and the efficiency of transmission of these vibrations.

9.7.2   Measurement, Vibration or Noise?

The choice between noise monitoring and vibration monitoring must be decided for each situation considered. Noise levels are often more convenient to measure since no instrument need be attached to the machine. However extraneous noise can cause problems of masking. Vibration measurement is more selective and inherently more repeatable, and for this reason is generally used in preference to noise monitoring. The transducer is placed on or near the part of the machine that is to be checked. For instance, the transducer would be placed on the bearing housing if the bearing condition were to be checked and on the pump casing if cavitation were of interest. Most vibrations are associated with moving mechanical parts. Consequently the bearing housing will be the appropriate location for most measurements.

9.7.3   The Equipment

The noise transducer is a microphone which converts acoustic pressure signals into electrical signals. The conversion device is commonly a piezoelectric element or a capacitive element coupled to a pressure sensitive diaphragm. The most common vibration transducer, the piezoelectric accelerometer, features small size, robust construction, and wide frequency and dynamic ranges. Moving coil vibration transducers

174

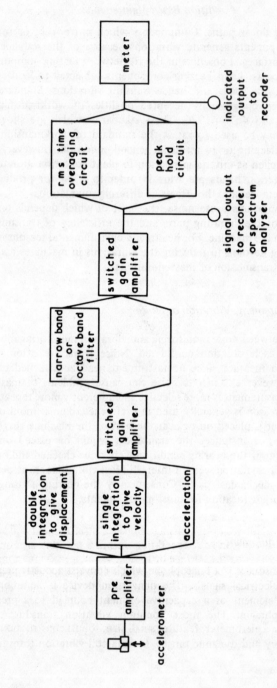

Figure 9.4. Schematic diagram of a vibration meter

sensing velocity are also used, but they tend to be both heavy and bulky. Such transducers were the first to be applied to vibration monitoring but they have now been superseded by piezoelectric accelerometers for most applications. The chief drawback of any transducer operating by the piezoelectric effect is that the motion is transduced into a charge which is then dissipated in the signal conditioning equipment and the cable. If the rate of change of the motion is slow this dissipation affects the reading significantly. Thus a low frequency limit is placed on the useful range of the device. However modern charge amplifiers, which have input impedances typically as high as 100 GΩ, enable measurements to be made at frequencies as low as 0.1 Hz.

All the transducers require signal conditioning equipment to convert the low level signal from the transducer to a signal in the region of 1 V r.m.s. Most commercial signal conditioning units also contain visual read-out, range changing and calibration facilities in addition to the basic signal amplifier. Integrators are incorporated to convert the acceleration signal to velocity or displacement. *Figure 9.4* is a schematic diagram of a vibration meter. More detailed information about transducers and noise and vibration meters may be found in Refs. 10, 11 and 12.

9.7.4 The Vibration or Noise Signals

Before discussing the various techniques of signal monitoring it is necessary to examine the typical features of a vibration or noise signal and the ways by which these features are described.

The signal fluctuates with time about a mean level which is often but not always zero. The pressure due to noise is superimposed upon, and fluctuates about, atmospheric pressure. An accelerometer mounted vertically is subjected to a steady gravitational acceleration in addition to the fluctuating vibration. In such cases the steady, or zero frequency, component is removed in the transducing or signal processing stages.

The signal could have one of the following forms:

1. Pure sinusoidal such as would be generated by an unbalanced rotor. This signal has only one frequency component, as shown in *Figure 9.5(a)*. Either the peak value or the root mean square value (r.m.s.) can be used to describe the size of the signal since for a sine wave they are directly related, i.e. $V_{peak} = \sqrt{2} V_{r.m.s.}$. The signal repeats itself with a period given by 1/frequency.

2. Periodic but non-sinusoidal containing many discrete frequency components, *Figure 9.5(b)*. The relation between the peak and r.m.s. values of the signal and the relative magnitudes of the different frequency components both change with the form of the signal. Such a signal

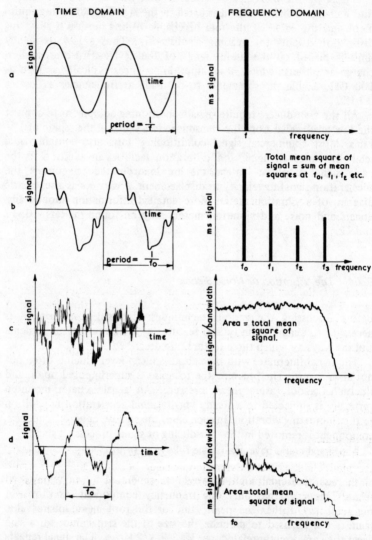

Figure 9.5. Typical noise or vibration signals and their frequency spectra

would be generated by the motion of the piston in an internal combustion engine and by the combustion forces. The repetition period of this signal corresponds to the lowest or fundamental frequency component and is given by $1/f_0$.

3. Random and never repeating itself exactly. Fluid flow generates such a signal, as does moving contact between solid surfaces. The peak value is of no interest since, in theory at least, it reaches infinity for infinitesimal periods. In practice any random signal has an upper frequency limit, which implies a cut-off in the peak size of the signal. Thus a random signal is described by its r.m.s. value and by its frequency spectrum, which is continuous over a wide frequency range (*Figure 9.5(c)*).

4. Any combination of sinusoidal, periodic and random such as the signal shown in *Figure 9.5(d)*, measured at the bearing of a small electric motor.

Thus any signal can be described by a measure of its magnitude, e.g. the r.m.s. time-average value, and a measure of its frequency spectrum.

The root-mean-square time-averaging parameter is the most widely accepted measure of the size of a signal. It is defined by the expression

$$V_{\text{r.m.s.}} = \sqrt{\left\{\frac{1}{T} \int_0^T V^2 \, dt\right\}} \qquad (9.2)$$

where V is the instantaneous value of the signal, which fluctuates about zero. The averaging period, T, must be much longer than the period of the lowest frequency component of the signal. Random signals with an energy spectrum extending to zero frequency therefore need an infinite time for accurate measurement. However, in vibration monitoring it is unusual to consider signals containing significant frequencies lower than 2 Hz and most types of vibration equipment incorporate filters to remove signals below this frequency. Hence, a realistic averaging time of about 2 s will permit measurements without undue fluctuation error. The errors that do occur manifest themselves as unsteadiness in the meter indication.

It is not merely fortuitous that r.m.s. time averaging is so widely accepted. The square of the vibration signal or the noise signal is a measure of the energy content of the signal. Kinetic energy is proportional to velocity squared, strain energy is proportional to displacement squared and acoustic energy is proportional to pressure squared. Thus the r.m.s. value of the signal reflects the mean energy value. It follows, therefore, that the total mean square value of a signal can be obtained by summing the mean square values of the various frequencies making up the signal, i.e.

$$(V_{\text{r.m.s.}}^{T})^2 = \sum_{n} (V_{\text{r.m.s.}}^{n})^2 \tag{9.3}$$

where n represents the appropriate component of the signal. For a periodic signal as shown in *Figure 9.5(b)* the mean square total motion is simply the sum of the mean square values of each frequency component. In the case of the random signal shown in *Figure 9.5(c)* the mean square total motion is given by the area under the frequency spectrum curve. This concept of addition of mean square components is important when frequency analysis is used in signal monitoring.

It is worthwhile noting at this point that unless the signal is a pure sine wave the time-averaging circuit must be a true mean squaring circuit and not a rectified mean circuit recalibrated to r.m.s., as is found in general purpose meters, e.g. the standard Avometer.

The measurement of the size of the signal is further complicated by the extensive use of logarithmic units in place of the more familiar linear units. Though confusing for persons who are not familiar with it this practice has several advantages. Natural and physical phenomena tend to be dependent on the ratios of values not the differences, e.g. 528 Hz has the same relation to 264 Hz (middle C) as 1056 Hz has to 528 Hz. That is they are both one octave apart. The use of logarithmic scales enables very large ranges of signal magnitudes to be compressed into small ranges of numbers, and in graphical presentations relative accuracy is maintained constant across the range.

The logarithmic scale in general use is the decibel scale which can be defined in terms of energy or power as

$$L_1 - L_2 = 10 \log_{10} (E_1/E_2) \tag{9.4a}$$

where L_1 is the signal level in dB corresponding to energy E_1 and L_2 is the level corresponding to E_2. Since energy is proportional to the square of the relevant vector quantity, V, e.g. displacement, velocity, or voltage, the above equation can be written as

$$L_1 - L_2 = 10 \log_{10} (V_1^2/V_2^2) = 20 \log_{10} (V_1/V_2) \tag{9.4b}$$

The decibel is thus a scale relating ratios of values. To be applied as a scale for absolute values, reference values, of the appropriate variables, at which $L = 0$ must be defined. Internationally defined reference levels are:

Acceleration	Velocity	Displacement	Sound pressure
10 μm/s	10 nm/s	none	20 μPa

Table 9.5 shows the additive nature of the decibel scale when used to compare signal values. An additional useful feature of the decibel scale, which stems from its definition (equations 9.4a and 9.4b), is that both the vector quantities and their associated energy quantities are represented by one logarithmic level, i.e. when the voltage ratio is 2 the corresponding power ratio is 4, both ratios being equivalent to a level difference of 6 dB.

Table 9.5. THE DECIBEL SCALE

Level difference, dB	Vector quantity ratio	Energy quantity ratio
0	1.0	1.0
1	1.12	1.26
2	1.26	1.6
3	1.4	2.0
5	1.73	3.0
6	2.0	4.0
10	3.16	10.0
20	10.0	100.0
60	10^3	10^6
120	10^6	10^{12}

The frequency spectrum can be described graphically (as shown in *Figure 9.8*) or numerically as a set of r.m.s. band-filtered measurements. The frequency spectrum of a machine is often called its *vibration signature*. The 'signatures' of two similar machines may not be similar. However they do tend to remain stable unless a change in the condition of the machines occurs.

9. 7. 5 Practical Vibration Monitoring Techniques

On-load monitoring as described in Section 9.2.1 is by far the most commonly used method of vibration or noise monitoring. In this and the ensuing sections the monitoring techniques described will apply specifically to vibration monitoring. However, most of these techniques are just as appropriate when monitoring noise. On-load monitoring can be performed in two main ways.

1. *Periodic field measurements with portable instruments.* This is the basic general purpose method which provides information about

Table 9.6. VIBRATION CHARACTERISTICS OF SOME MACHINE FAULTS

Fault	Dominant frequency	Direction position	Comments
1. Rotating imbalance	Rotation speed (n)	R B	Machinery running near or above a critical speed can change balance with speed – multiplane balancing is then required
2. Reciprocating imbalance	$1n$, $2n$, $3n$, $4n$ etc.	R B	Fundamental (n) may be balanced leaving only higher orders of imbalance
3. Mechanical looseness	$2n$, $3n$, etc.	R, A B	Affected by temperature and speed
4. Bent shaft or coupling	$1n$	R B	Can be caused by thermal effects. Often occurs with imbalance (1)
5. Misalignment	$1n$ or $2n$, sometimes $3n$, $4n$	R, A B	Common fault. Can be masked by characteristics of flexible couplings
6. Cracked shaft Asymmetric shaft	$1n$, $2n$, $3n$	R B	Transverse stiffness varies with direction and results in deflection under gravity exciting vibration. Best detected when machine is run-down through resonance. Deep crack has small effect
7. Damaged hydro-dynamic or hydrostatic bearing	Change in higher frequencies as debris causes rubbing	R B	Sometimes reduced clearance causes reduction in vibrations due to other effects
8. Journal bearing loose in shaft or housing	$n/2$, $n/3$	R B	Looseness affected by temperature or centrifugal effects
9. Oil film whirl	$0.4n$ to $0.5n$	R B	Oil properties (hence temperature) affect whirl

Fault	Symptom	Code	Notes
10. Hysteresis whirl	Shaft speeds above critical	R	Due to hysteristic damping in shaft. Excited when passing through critical speed and does not die away as speed increased. To cure reduce hysteresis or increase external damping, e.g. bearings
11. Damaged rolling element bearing	Input frequencies for bearing components and high frequencies in the range 500–5000 Hz and shock pulse detectable at very high frequencies, 10–100 kHz	R, A B	Impact frequencies for rolling element dia. d, pitch dia. D, contact angle, β, and bearing speed, f. Defect in one ball $f_b = \dfrac{fD}{d}\ \left(1 - \left(\dfrac{d\cos\beta}{D}\right)^2\right)$ Defect in $\begin{array}{c}\text{inner}\\\text{outer}\end{array}$ race $f_r = \dfrac{fn}{2}\left(1 \pm \dfrac{d\cos\beta}{D}\right)$ Roller bearing, $\beta = 0$, thrust ball, $\beta = 90°$. Taper roller, d is dia. of roller at pitch dia., D, and β = inclination of roller axis
12. Damaged, worn or misaligned gears	Tooth meshing frequency + harmonics shocks	R, A B	Sideband frequencies, detectable with very narrow band analysers, indicate gear eccentricity. Increased peak to r.m.s. ratio indicates damaged teeth
13. Damaged belt or chain drive	Multiple of belt or chain speed	R B	
14. Cavitation or turbulence	Increase in mid to high frequencies, 0.5–10 kHz	C	Indicates, wrong flow, passage block or passage damage
15. Damaged blades, vanes, guides, tubes, etc.	Increase in frequencies associated with blade passing or vortex shedding	C	Usually designed to avoid worst frequencies but damage may induce flow fluctuations
16. Electrical faults	Multiples of supply frequency	C B	Burnt out phase induces cycling forces. Can also cause shaft imbalance

R = Radial,   A = axial,   B = bearing housing,   C = casing long-term changes in the condition of plant. The portable instruments are employed with a high load factor and can often be placed in the care of only one man. The interval between measurements must be determined by experience. If the machinery under consideration breaks down frequently the interval between inspections should be short, perhaps a week. On the other hand for plant with infrequent failures, weekly monitoring would be wasteful and perhaps monthly would be more reasonable. In any event the frequency of monitoring can be adjusted to fit the operational experience of each type of machine in relation to frequency of breakdown, severity of the effects of a breakdown, and the degree of warning provided by the particular technique of monitoring being applied. Use of life curves and the LEO approach (Sections 9.4.2 and 9.4.3 will assist decision making).

2. *Continuous monitoring with permanently installed instruments.* Continuous monitoring is employed when machine failures are known to occur very rapidly and/or when the results of such a failure are totally unacceptable as is the case, for example, with turbine generator units.

Locating the monitor points and mounting the transducers Location of the transducers is dependent on the type of malfunction to be kept under surveillance. *Table 9.6* gives some guidance on this. It will be seen that the bearings, being directly connected to the moving parts, provide the most useful monitoring positions. In the case of vibration transducers the orientation of the transducer is important. An axially placed transducer will pick up vibrations due to bearing noise whilst being less sensitive to shaft imbalance than either vertical or horizontal transversely placed units. When machines are mounted directly on the factory floor, the flexibility of the floor in the vertical direction allows transmission of vertical vibrations from one machine to another. In these situations horizontal vibration measurements will tend to be more meaningful. On the other hand, if the machinery is mounted several floors up in a space-frame building horizontal measurements may be affected by other machinery mounted on the structure.

It is important to locate vibration transducers on a machine member that will transmit the pertinent vibrations to the transducer. Since all materials have mass, and are elastic, every component will exhibit a natural frequency at which it will vibrate with increased magnitude in response to being forced. The response of the component to forcing at different frequencies will be as shown in a simplified form in *Figure 9.6*. From this figure it will be seen that the component will not respond significantly to forcing at a frequency well above its natural frequency.

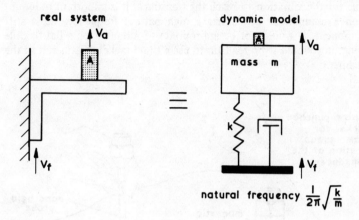

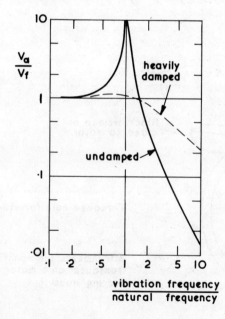

Figure 9.6. Response of a simple elastic system (A represents the transducer)

Thus for the vibration to reach the transducer it is important to mount it on a component which has a high natural frequency, i.e. a stiff component. It would not be appropriate to mount it on a flat flexible casing, nor would it be realistic to use a thin bracket to attach it to the machine.

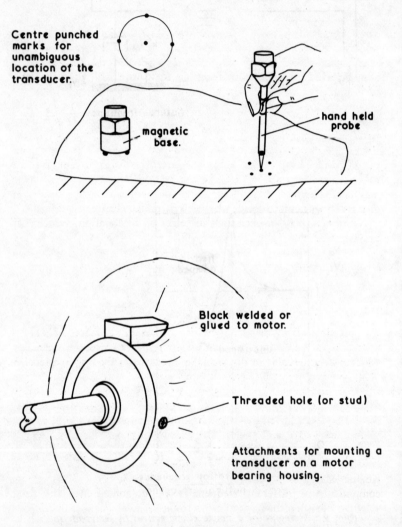

Figure 9.7. Methods of locating or mounting the transducer

Since the vibration level will vary with transducer location the mounting point should be clearly and unambiguously marked. Centre punched marks as shown in *Figure 9.7* are appropriate for use with pointer, magnet, or lever grip mounted transducers. For more repeatable readings the use of a threaded hole, or a welded glued block, or stud, is satisfactory.

Total signal monitoring This is the most straightforward method, resulting in only one measurement for each transducer location. The total signal picked up by the transducer is amplified, time-averaged and displayed on a meter or recorded for future use. The schematic of a typical set of equipment is shown in *Figure 9.4*. Such equipment is compact, being typically contained in a box about 200 mm cube and weighing only 2.5 kg. Broch[10] describes the application of typical portable equipment to total signal monitoring. The vibration reading so obtained is noted on the machine records, which should be clearly laid out in tabular form or alternatively in graphical form. A change in the measurement between inspections indicates that some machine change may have occurred. Until experience with the machine has been built up it is very difficult to decide whether a particular change is significant.

Vibrations may be measured in terms of acceleration, velocity or displacement. For a given frequency, f,

$$(\text{velocity})_{\text{r.m.s.}} = 2 \pi f \times (\text{displacement})_{\text{r.m.s.}}$$

and

$$(\text{acceleration})_{\text{r.m.s.}} = 2 \pi f \times (\text{velocity})_{\text{r.m.s.}} \tag{9.5}$$

The practical effect of these relations is that measurement of acceleration accentuates the high frequencies with respect to the low frequencies, whereas measurement of displacement accentuates the low frequencies. Many vibration meters (*Figure 9.4*) provide for the acceleration output of the transducer to be integrated once or twice to generate velocity or displacement. *Figure 9.8* shows acceleration, velocity and displacement frequency spectra of the vibrations of a motor bearing housing, plotted on logarithmic scales. The accentuation provided by each of the three types of measurement is clearly visible. R.M.S. velocity is considered to be the most useful measure for total vibration monitoring. Kinetic energy is proportional to mass $\times$ (velocity)^2 so vibrational components at different frequencies are combined into the total velocity measurement in proportion to their kinetic energies unaffected by frequency. BS 4675[9] recommends r.m.s. velocity measurement in the frequency range 10–1000 Hz.

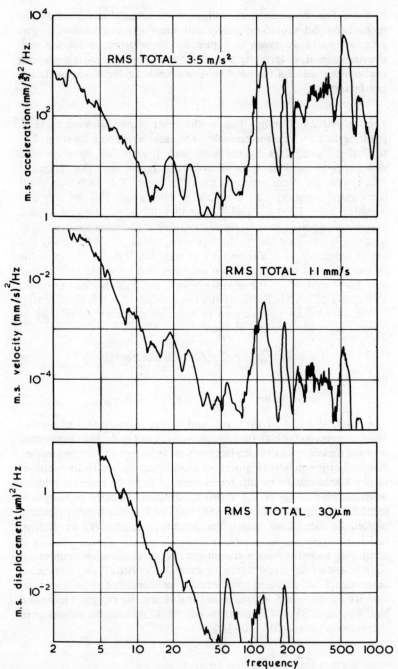

Figure 9.8. Frequency spectra of the vibrations at a motor bearing housing

With noise the situation is somewhat similar. Sound pressure is proportional to local sound velocity and hence is a measure of energy in the sound unweighted by frequency. Most sound level meters incorporate a switchable 'A weighting' network which attenuates the low frequencies and accentuates the middle to high frequencies in order to simulate the characteristics of human hearing.

As already discussed in Section 9.4.1 keeping records is essential in any monitoring programme since its purpose is to detect changes. Nicholls[15] describes suitable record sheets. In general it is important to record on a standard form the machine details, locations and types of transducer, axis of measurement (vertical, transverse or axial), the units of the measurements, any frequency limits, magnitudes of changes at which it is proposed that action be taken, data and operator. A space for a sketch of the machine, possibly including a photograph, is very useful.

When continuous monitoring is employed the processed signal will be displayed on a chart recorder. It is normal to arrange the equipment with alarms set to trigger at a predetermined signal level, change in level or rate of change in level[11]. Modern turbine-generator units have vibration transducers mounted vertically on each bearing pedestal. The signal levels are recorded on multipoint chart recorders which actuate alarms in the control room if any one pedestal has reached a predetermined vibration level.

Automatic recording can be used to produce graphical records, either continuous or scanned, and computer control has been used in large schemes. However, for the bulk of monitoring schemes hand measurements and hand kept records are quite adequate.

Frequency analysis In many situations a change in the vibration of one frequency component, which is indicative of trouble, can be masked by a more dominant but acceptable vibration at another frequency. In such circumstances frequency analysis is essential.

When the frequency of vibration associated with a maintenance problem is known in advance frequency analysis provides additional useful indication of the cause. *Table 9.6* lists the more common machine faults and their vibration characteristics.

In the frequency analysis process an electrical filter network is employed to reject part of the frequency spectrum whilst allowing another part to pass through the filter. Four types of filter are common, low pass, high pass, band pass and band reject (*Figure 9.9*). The band pass filter is used in frequency analysis although the other filter types have uses in specific situations. It is not possible to produce a filter with the ideal sharp cut-off shown by broken lines in the figure. All practical filters have sloping characteristics such as given by the solid lines. The bandwidth of a filter with sloping characteristics is normally quoted as its 'half power' bandwidth, i.e. the frequency bandwidth between the points where the voltage transmission coefficient is $1/\sqrt{2}$.

Practical spectrum analysers fall into two types. Those which maintain a constant bandwidth across the frequency spectrum are shown in

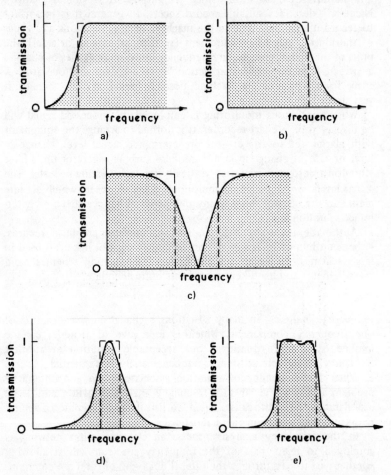

Figure 9.9. Frequency transmission characteristics of types of filters. (a) High pass, (b) low pass, (c) band reject, (d) band pass (narrow band), (e) band pass (broad band). Broken lines – ideal; solid lines – real

Figure 9.10 (top) and those with a bandwidth which is proportional to the frequency to which the filter is tuned, in *Figure 9.10* (bottom). The constant proportional bandwidth analyser has the effect of enhancing the higher frequencies. Many machine faults generate periodic signals (*Table 9.6*). The use of a filter suppresses the signal from random vibrations in comparison with periodic signals, and is therefore very useful for monitoring machine faults.

The bandwidths employed in practical frequency analysers are within the range 1% and 70% (1 octave). The narrow bandwidths are very selective but analysing times will be long since many measurements are

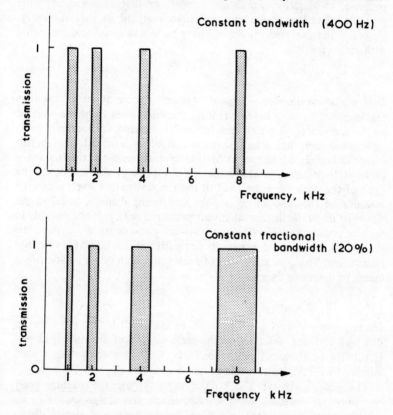

Figure 9.10. Comparison of the pass band characteristics of frequency filters. (When plotted on a logarithmic frequency scale, the compression of the scale as frequency increases results in the constant fractional bandwidth filter appearing to have constant bandwidth, whilst the true constant bandwidth filter appears distorted in width)

required to cover the full frequency range of interest. The widest band-width (1 octave) is less selective but rapid in use. In field analysis using portable instruments, 1 octave or 1/3 octave switched filtering is normally quite adequate. Such equipment is readily obtained in compact form from a range of suppliers[9]. The narrower bandwidths are usually obtained using swept frequency filters in which the filter can be centred on any chosen frequency. Automatic analysers which sweep the frequency range and plot out the spectra, and real-time analysers which monitor up to 500 bands virtually simultaneously, are available to reduce labour in monitoring. Such equipment could hardly be called portable, so vibration signals are recorded on magnetic tape using the portable vibration equipment and then analysed in the laboratory. However, the analysed results still have to be examined and compared with earlier results.

Peak signal monitoring — figure of merit Some types of machine malfunction, such as bearing failure, cause impulses to be transmitted to the transducer. A pitted race in a rolling bearing, for example, emits an impulse each time a ball comes in contact with the pit. Though this causes an increase in the r.m.s. vibration level, especially at the frequency of impacts, it has a greater effect on the peak of the vibration. The ratio of the peak vibration level to the r.m.s. vibration level, which is a measure of the amount of impulse generating damage, is called the *figure of merit*. With normal vibration instruments, it is not possible to register peak values of vibrations containing transients of duration less than 10 μs. However, this presents no significant restriction for vibration monitoring. Thus, measurement of peak vibration levels is an additional means of detecting damage.

Shock pulse monitoring[16] Impulses emanating from impacts between damaged surfaces, described in the previous section, are employed very effectively by the shock pulse monitoring method which was developed initially by S.P.M. Instruments Ltd, of Sweden.

The impacts generate shock pulse waves through the machine body and they are detected using a piezoelectric transducer similar to an accelerometer except that it is tuned mechanically and electrically to have a resonant frequency at 32 kHz. The transducer resonates at this frequency in response to each pulse (*Figure 9.11(a)*). The signal generated by the transducer is filtered to separate the resonant frequency from the background of vibrations. The magnitude of each resonance is held

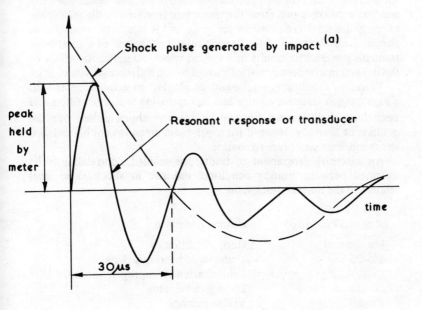

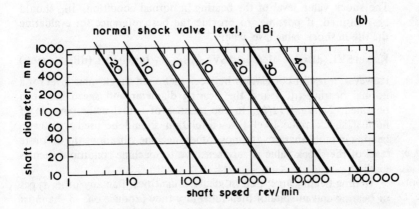

Figure 9.11. Shock pulse monitoring. (a) Response of transducer signal to shock pulse generated by impact; (b) relation between normal shock valve level and diameter and speed of a normally loaded bearing

for sufficient time to be read on the meter. By this means the peak amplitude of the signal from the transducer is related to the magnitude of the pulse wave generated in the body and is relatively insensitive to the mechanical design of the bearing housing. The use of a transducer resonating at 32 kHz results in a system which magnifies the effect of the impacts in a frequency not influenced by general machine vibrations.

Vibration monitoring equipment employing an accelerometer with a high enough frequency range and incorporating frequency filters and peak hold facilities can be used to measure shock pulses, but the equipment specially designed for shock pulse measurement has reduced the monitoring procedure to routine.

An extensive programme of testing has enabled a correlation to be obtained between bearing condition and rise in shock value level relative to the normal shock value.

Rise in SVL, (dB_N)	*Bearing condition*
less than 20	Good condition
15–25	Undamaged bearing lacking lubrication or improperly installed
20–35	Damage developing
35–50	Visible damage
50–60	Risk of breakdown

The shock value level of the bearing in normal condition, dB_i, should be measured, if possible, to provide the best reference for evaluating the rise in shock value level where

$$\text{Rise in SVL}\,(dB_N) = \text{Measured SVL}\,(dB_{SV}) - \text{Initial SVL}\,(dB_i) \quad (9.6)$$

However, a useful correlation between measured shock value level for a normal bearing (dB_i) and the bearing diameter and speed has been obtained, *Figure 9.11(b)*. In the absence of actual measurements of normal shock value level this correlation should be used with care, bearing in mind that as with most other machine vibrations the absolute level of the shock value is very sensitive to machine conditions such as loading.

Rolling bearings are used in greater quantities than any other types of bearing and in spite of their reliability they produce one of the most common types of failure in industrial plant. The shock pulse method has been thoroughly developed and is now widely used for monitoring. Of course the technique is capable of detecting many other sources of impact, such as damage or misalignment in transmission systems, but experience of measurements on such systems is relatively limited.

Monitoring of large installations Where plant costs are high and shutdown is expensive computer monitoring can be employed. In large installations containing many machines, continuous or scanned monitoring of transducers permanently mounted on the machines is being used to assist communications and maintenance planning. Such systems can incorporate malfunction alarms and displays on each machine together with remote indication at a central real-time maintenance computer[17]. To assist diagnosis and treatment of a machine malfunction the computer can be programmed to list a machine's history, susceptibility to types of malfunction, typical and previous symptoms, repair work sheets, and material order forms etc. Such sophisticated systems tend to reduce maintenance problems on the primary machines at the expense of introducing a maintenance problem in the monitoring instrumentation. Nevertheless they may be economic in certain situations. However, the vast majority of monitoring problems can be handled adequately using portable vibration monitoring equipment incorporating octave or 1/3 octave manually switched filters.

Special vibration monitoring techniques The methods that have been described are adequate for monitoring most plant. They involve simple equipment and could be undertaken by any practising engineer. However there are more sophisticated techniques which, while increasing the engineer's ability to detect and diagnose the early stages of plant damage, do demand a greater understanding of vibration and spectral analysis. Typical of these are frequency tracking, run-down tests[18, 19], correlation analysis[10, 20], time domain analysis[12], statistical amplitude analysis[20], stresswave emission[21], and Cepstrum analysis[22, 23]. Detailed descriptions may be found in the references quoted.

9.8 A Programme for Plant Condition Monitoring

Although this focuses on vibration monitoring the approach is just as relevant to other types of monitoring.

Select the machines to be monitored Examine the plant to see if a plant condition monitoring problem exists. The frequency, rate of progression, and effects of plant breakdown will be the main criteria. It is not wise to include too many machines initially. It is better to wait until experience of monitoring a few machines has been built up before expanding the programme. One or more machines should be selected which have a bad maintenance record due to failures which could be expected to give advance warning if monitored.

Determine the type of monitoring required   Long term monitoring is provided by periodic checks. Immediate warning is provided by continuous monitoring. It is not normally possible at this stage to decide whether frequency analysis is necessary or whether total signal measurements will be sufficient.

Train an engineer to conduct the monitoring programme and select suitable instrumentation   Sophisticated instrumentation should be avoided unless the problem demands it. A simple portable vibration and/or noise meter with a manually operated frequency filter will enable a useful monitoring programme to be set up in most situations. Even so the initial equipment including the transducer and a filter unit will cost in the region of £1000 (1976 prices). A wise precaution is to choose instrumentation that can be 'expanded' at a later date even though this may restrict future purchases to the products of the original equipment supplier. The better instrument suppliers are experts in their fields and will advise on particular problems.

The responsibility for the monitoring measurements should rest with one engineer who can be trained, and who will then gain experience in the use of the equipment and in the interpretation of the measurements.

Select the location of the monitor points and choose the interval between periodic checks   For each machine decide what types of machine malfunctions are to be kept under surveillance. Then select and mark locations for the transducers appropriate to these malfunctions. *Table 9.6* gives some guidance on this.

The interval between checks is chosen on the basis of operational experience with each machine, using life curves as described in Section 9.4.2. However, administratively there is an advantage in keeping the intervals the same for every machine in the factory.

Determine the 'normal' conditions of the machine   Initial vibration measurements will be carried out and examined. Frequency analysis is useful at this stage in order to identify sources of vibration and to establish the machine 'vibration signatures'. Subsequent measurements must be carried out before the 'normal' condition of each machine can be established with any confidence.

The mechanical condition of the machine is clearly relevant to the normal vibration levels. During this process it may be found possible to reduce the number of readings taken on any one machine by excluding certain frequency bands, or some measurement point, from the monitoring.

This is the appropriate stage to draw up a suitable vibration history form to enable subsequent measurements to be related to existing data.

Action in the event of abnormal vibrations being monitored The final problem is to decide what deviation from the normal signature is to be considered significant enough to warrant the shutdown of the machine for maintenance or repairs. The first action on obtaining abnormal results would be to repeat the measurements and endeavour to pinpoint the source of the change of vibration. If changes in vibration levels are fitted to exponential time or similar curves it becomes possible to predict the rate of deterioration (Section 9.4.2).

As experience builds up and monitored vibration levels have been obtained, leading up to breakdown on a machine or batch of machines, it will become possible to set vibration level envelopes. If the monitored vibration level moves outside its set envelope the unit would be scheduled for maintenance with a priority based on previous experience of the rate at which the machine can be expected to deteriorate towards actual breakdown. These envelopes would usually be upper limits but in certain cases lower limits would also be appropriate.

Whether experience is limited or not, once the cause of a monitored change in the vibration level is known the decision to shut down can be made on the same basis as any other management decision; however, when condition monitoring is being employed the diagnosis, and hence the managerial decision is, at least in part, scientifically based.

REFERENCES

1. Neale, M. J. and Woodley, B. J., 'Condition Monitoring Methods and Economics', *SEECO 75,* Imperial College, 23–24 Sept. (1975)
2. Birchon, D., 'Non-destructive Testing', *Engineering Design Guide 09,* OUP (1975)
3. Birchon, D., 'Some Diagnostic Techniques in Machinery Health Monitoring', Paper C97/73, *Conf. on Machinery Health Monitoring,* I. Mech. E. (1973)
4. Scott, D., 'Failure Diagnosis and Investigation', *Tribology* 3, 1, 22–28 (1970)

5. Scott, D., 'Debris Examination – a Prognostic Approach to Failure Prevention', *Proc. 8th Israel Conf. on Mech. Eng. Wear* (1975)
6. BS 3889, *Methods for Non-destructive Testing of Pipes and Tubes,* BSI, London (1965)
7. Birchon, D., 'The LEO Technique', *Engineer,* 226, 478–481, London (1968)
8. Birchon, D., 'The LEO Technique', *Br. Jn. Non-destr. Test.,* 12, 3, 73–78 (1970)
9. BS 4675, *'Mechanical Vibration in Rotating and Reciprocating Machines,* Part 1, *Basis for Specifying Evaluation Standard for Rotating Machines with Operating Speeds from 10 to 200 Revolutions per Second',* BSI, London (1976). Also, ISO Rec. 2372:1974, VDI 2056, and DIN 45665.
10. Broch, J. T., *Mechanical Vibration and Shock Measurement,* B & K Laboratories Ltd. (1972)
11. B & K Application Note 14-227, *Notes on the Use of Vibration Measurement for Machine Condition Monitoring*
12. Randall, R. B., 'Vibration Signature Analysis – Techniques and Instrumentation Systems', *Noise, Shock and Vibration Conference,* Monash University (1974). Also B & K Application Note 14-138
13. Collacott, R. A., 'Component Life Concepts Related to a Theory of Whole-life Expectancy', *Quality Ass.* 1, 4, Dec. (1975)
14. Sankar, T. S. and Xistris, G. D., 'Failure Prediction through Theory of Stochastic Excursions of Extreme Vibration Amplitudes', *ASME Paper 71-VIBR-60* (1971)
15. Nicholls, C., 'Vibration Monitoring and Analysis of Critical Machinery', *The Plant Engr.,* 13, June (1974)
16. Brown, P. J., 'Preventive Maintenance of Machinery Bearings', *Maintenance Engineering,* Sept. (1976)
17. Johanson, K. E., 'Maintenance System for the Mechanical Industry including Automatic Condition Monitoring System and Administrative System for Maintenance', *Report LiTH-IKP-R-R-071,* Linkoping Inst. of Tech., Department Engineering, Sweden (1976)
18. Henry, T. A. and Okah-Avae, B. E., 'Vibrations in Cracked Shafts', Paper C162/76, *Proc. I. Mech. E. Conf.* Vibrations in Rotating Machinery, Cambridge, Sept. (1976)
19. Mayes, I. W. and Davies, W. G. R., 'The Vibrational Behaviour of a Rotating Shaft System Containing a Transverse Crack', Paper C168/76, *Proc. I. Mech. E. Conf.,* Vibrations in Rotating Machinery, Cambridge, Sept. (1976)
20. Macleod, I. D. *et al.,* 'Analysis Techniques for Use in Acoustic Diagnostics', *Conf. on Acoustics as a Diagnostics Tool,* 81–102, I. Mech. E., London (1970)
21. Arrington, M., *Acoustic Emission, an Introduction for Engineers',* I. Mech. E. CME, 53–55, April (1975)
22. Randall, R. B., 'Cepstrum Analysis and Gear Box Fault Diagnosis', *B & K Application Note,* 13–150
23. White, K. J., 'Detection of Gear Box Failures', *Workshop in On-condition Maintenance,* I.S.V.R., Southampton, Jan. (1972)

Chapter 10

Management Techniques in Maintenance

Of the various management techniques that can contribute to the productivity and effectiveness of the maintenance department the most important are method study, work measurement, organisation and methods, activity sampling and logical fault finding.

10.1 Method Study

This is one of the two major techniques of work study, work measurement being the other (see *Figure 10.1*). It is defined[1] as the systematic recording and analysis of existing and proposed ways of doing work, as a means of developing easier and more effective methods. *Its standard procedure (see Figure 10.2) is simply a systemisation of the logical procedure for any investigation.*

In any organisation the resources available for method study are limited and attention must therefore be focused on those areas that are most likely to yield the greatest net cost reduction. A system for the identification of such areas was outlined in *Figure 1.4* where it was shown that the causes of high maintenance cost might be technical (i.e. poor design) or organisational (i.e. poor work methods). Thus method study (or its extension, organisation and methods) can be applied to maintenance in a number of ways, i.e.

as an aid to designing out maintenance   technical

improving documentation   organisational
methodising jobs
improving organisation and planning

While the general procedure of method study is always followed the techniques appropriate for recording and analysing the facts vary from problem to problem.

WORK STUDY

'The generic term for those techniques, particularly work measurement and method study, which are used in the examination of human work in all its contexts, and which lead systematically to the investigation of all the factors which affect the efficiency and economy of the situation'

METHOD STUDY

'The systematic recording and examination of existing and proposed ways of doing work, as a means of developing and applying easier and more effective methods'

WORK MEASUREMENT

'The application of techniques designed to establish the time (standard time) for a qualified worker to carry out a specified job at a defined level of performance'

Figure 10.1. The major techniques of work study

SELECT	the job or problem
DEFINE	the objective
RECORD	all the relevant facts
EXAMINE	critically all the activities
DEVELOP	the best method
INSTALL	the agreed method
MAINTAIN	the modified method

Figure 10.2. The seven-step method study approach

The *selection* step of a maintenance method study is partially incorporated in the maintenance control system mentioned above. Such a control system should identify, classify and arrange in order of priority those high cost maintenance areas which are suitable for solution by some form of method study. Technical studies would usually be the responsibility of the maintenance or engineering services; organisational studies might well be carried out by a specialist method study team. In both cases the initial step should be a pilot study to establish the feasibility of the project.

Before any evaluation of a situation the necessary information must be collected for *recording* in the most convenient form. The technique of recording should be that which is best suited to the circumstances. Job methodisation studies might use detailed charts of the type shown in *Figure 10.3*; maintenance organisational studies

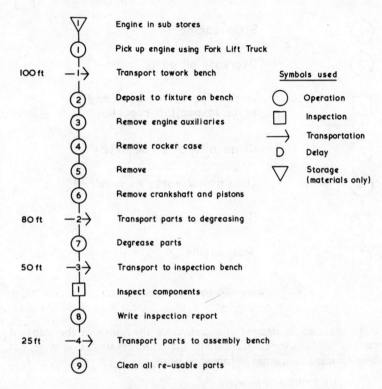

Figure 10.3. Flow process chart

might use the less detailed flow process chart, as shown in *Figure 10.4*. *Figure 10.5* illustrates how documentation procedures can likewise be represented. At a different level a whole maintenance system can be displayed as in *Figure 13.1*.

Critical examination of the information is the most important stage of method study. The job as a whole and, successively, its identifiable elements are examined using the procedure shown on the standard proforma, as in *Figure 10.6*, to determine whether — through elimination, simplification or modification — better method may be *developed*.

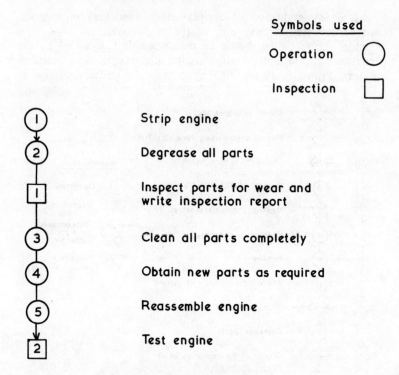

Figure 10.4. Outline process chart

In the case of *design-out maintenance* the whole of the standard method study procedure, including the critical examination stage, has been itemised on a series of forms[2] as below:

    (i) Outline of problem
   (ii) Elimination of unit (or assembly or sub-assembly to which troublesome part belongs)
  (iii) Substitution of unit
  (iv) Modification of unit
        (Identification of *cause* and possible solutions)
(v) – (ix) Installation, trial and financial justification of alternatives.

A procedure of this type can certainly assist maintenance engineers to sift the numerous alternative solutions to a maintenance problem. However, *before* such a procedure can be used the *cause* of the problem must be determined.

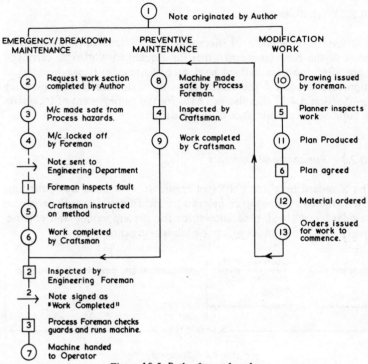

Figure 10.5. Path of a work order

CRITICAL EXAMINATION			
Description of Activity		Date	
		Ref	
Present Facts		Alternatives	Develop
WHAT?	WHY?	WHAT ELSE?	WHAT SHOULD?
HOW?	WHY?	HOW ELSE?	HOW SHOULD?
WHEN?	WHY?	WHEN ELSE?	WHEN SHOULD?
WHERE?	WHY?	WHERE ELSE?	WHERE SHOULD?
WHO?	WHY?	WHO ELSE?	WHO SHOULD?

Figure 10.6. Standard proforma for critical examination

10.2　Work Measurement

This has been defined[1] as the application of techniques designed to establish the time (standard time) for a qualified worker to carry out a specified job at a defined level of performance. It can be used to improve and control the effectiveness of maintenance work planning (see Section 5.4.3) and also as a basis for the employment of incentives to improve the performance of the tradeforce.

10.2.1　Fundamental Concepts

The 'standard time' for a job (see *Figure 10.7*) is made up of the time taken (the basic time) to complete a job, when working at the accepted 'standard rate', and time allowances for the ergonomic and environmental conditions under which the job is being carried out.

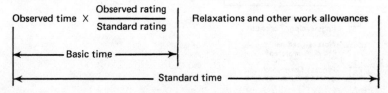

Figure 10.7. Standard time

The term 'standard rate' needs careful definition. Different people work at different rates and the times for identical jobs vary considerably. The British Standards Institution has its own scale, (there are others) against which a subjective estimate of a worker's rate can be measured. The most important point on this scale is the 'standard rate' which is defined as the rating corresponding to the average rate at which qualified workers will naturally work at a job provided they are *motivated* to apply themselves.

If the standard rating is maintained and the appropriate relaxation taken a worker will achieve *standard performance* over the working day. It is this ability to measure a worker's performance that provides the basis for incentive schemes.

10.2.2　Work Measurement Procedures

These are concerned with using data obtained from past or present observations to make reliable predictions about the times (standard

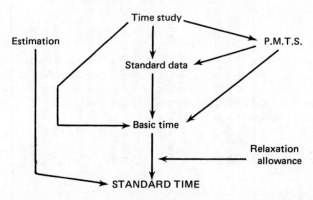

Figure 10.8. Work measurement procedures for obtaining standard time

times) that future jobs should take. Generally (see *Figure 10.8*) the standard times are based on data obtained by:

1. Direct observation, e.g. time study.
2. Synthetic methods.
3. Estimating from records or experience.

Time study   The standard time for a job is obtained by 'timing' and 'rating' the job as it is being carried out. In order to facilitate this the job is divided into easily distinguishable 'job elements'.

The basic time for an element is obtained by modifying the observed time to allow for the rate at which the element was judged to be performed (see *Figure 10.7*). The standard time is then obtained by adding the appropriate allowances to this basic time. The standard time for the complete job is then deduced by summing the standard times of each of its constituent elements.

Synthesis   Many job elements are common to a range of activities and their basic elemental times can be compiled into a library of 'standard data' from which standard complete job times can be synthesised without recourse to actual timing. Standard data can be compiled directly from time study or less directly from *predetermined motion time systems* (PMTS).

PMTS   Several systems are in use but only two, which are particularly relevant to maintenance work, will be described.

Table 10.1. PART OF THE MTM1 REACH TABLE OF TIME ELEMENTS

Distance moved, in	Time TMU				Hand in motion		Case and description
	A	B	C or D	E	A	B	
3/4 or less	2.0	2.0	2.0	2.0	1.6	1.6	A. Reach to object in fixed location, or to object in other hand or on which other hand rests
1	2.5	2.5	3.6	2.4	2.3	2.3	B. Reach to single object in location which may vary slightly from cycle to cycle
2	4.0	4.0	5.9	3.8	3.5	2.7	C. Reach to object jumbled with other objects in a group so that search and select occur
3	5.3	5.3	7.3	5.3	4.5	3.6	
4	6.1	6.4	8.4	6.8	4.9	4.3	D. Reach to a very small object or where accurate grasp is required
5	6.5	7.8	9.4	7.4	5.3	5.0	
etc							E. Reach to indefinite location to get hand in position for body balance or next motion or out of way
30	17.5	25.8	26.7	22.9	15.3	23.2	

Element analysis chart — Code 0720.02

Description Left Hand	No.	L.H.	T.M.U.	R.H.	No.	Description Right Hand
A Skin end of wire with knife # 10 or smaller						A
Move to area		MIOB	18·7	MI6C		Move knife to work
			16·2	P2SE		Align
			2·9	MIB		
			16.2	API		
			5.4	T9OS		Skin wire
			16.2	API		
			7.5	D2E		
			13·4	MI2B	·	Move knife away
			96·5			

Operation synthesis — Code 0720.02

Symbol	Ref.	Operation or element description	T.M.U.	Freq.	Total
720.0211		Skin wire # 10,12 and smaller (Levelled hrs. .0023)			
	05.0004	Handle knife			100.5
	04.0001	Handle wire			34.2
	A	Skin end of wire			96.5
					231.2

Universal standard data — Code 0720.02
Skinning

Symbol	Skin electrical conductor # 18 thru MCM			Hours
720.0211	Single	Small	# 10, 12 gauge and smaller	.0023
720.0212	conductor	Medium	# 4, 6, 8 gauge	.0034
720.0213	cable	Large	# 2 thru MCM gauge	.0076

Bench mark analysis sheet — Code 0720.02

Description : Medium size junction box — Date 26.4.61 — BM 0790 3
4 holes, 85 wires # 12, crimped — Craft. elect. Gen. install.
connections, mount and connect — Dwgs : None
No. of men: 1 — Analyst W.M. — Sht 1 of 1

Line	Men	Operation description	Reference Symbol	Unit time	Freq.	Total time
1		Mount medium size box	750.0207			·3243
2		Select proper wire	13.0002	.0035	85	.2975
3		Move marker on wire	720.0660	.0094	85	.7990
4		Cut off 85 # 12 wires	720.0101	.0021	85	.1785
5		Skin 85 # 12 wires	720.0211	.0023	85	.1955
6		Connect 85 # 12 wires	720.0323	.0110	85	.9350
7						
30						
Notes :			Bench mark time			2.7298
			Standard work group			G

Spread sheet task area, general insulation — Code 0795 — Craft : Electrical

Group E 1·2		Group F 2·0		Group G 3·0
1·5		2·5		
0790-16-Conduit, 15'-1¼" 2–30° bends, 2 condulets, 2 nipples between junction boxes. prepare conduit and install; 2 men. 0790-2 Medium size junction box. 4 holes, 37 wires.		0790 15 Conduit 35' 2" 2 30° bends, 2 condulets, 2 nipples between junction boxes. prepare conduit and install.		0790-3 Medium size junction box, 4 holes 85 wires # 12 crimped connections, mount and connect.

Figure 10.9. The build a spread sheet from MTM–1 using standard data

Methods Time Measurement 1. MTM1 is a system in which the standardised time data for basic human movements is tabulated in time units (TMU) of 0.000 01 h at a BS rating of 83.3. There are ten tables of standardised data covering the following movements: reach, move, turn and apply pressure, grasp, position, release, disengage, eye-travel, body motions, simultaneous motions. The reach table is shown in *Table 10.1.* In the case of highly repetitive short-cycle production work such tables can be used to establish the standard time directly. They can also be used to build standard data elements for maintenance work (e.g. Universal Maintenance Standards, UMS, employing much larger time elements than MTM). This is shown in the top three cards of *Figure 10.9.*

Methods Time Measurement 2. MTM2 is a simplified MTM system using a total of only 39 time values (c.f. the 120 time values in the reach table alone of MTM1). This simplified system has been used to build a specifically maintenance MTM system called Data Block Synthesis[3].

The techniques outlined were designed for short-cycle repetitive production work. In the case of maintenance work its *variability, diversity,* and *singularity* prohibit the use of such techniques. Some form of estimation is necessitated.

10.2.3   Maintenance Work Measurement

Work measurement techniques, based on the analytical techniques of Section 10.2.2, but also involving some degree of estimation, have evolved over the last few years. They are listed in *Figure 10.10.*

The most sophisticated of these techniques, *comparative estimating* based on Universal Maintenance Standards, was developed in the early 1950s by the Methods Engineering Council of Pittsburgh, U.S.A., as a yardstick for use in the introduction of incentives in maintenance. Its first U.K. application was in 1964 at the Leigh Works of BICC Ltd. (see Chapter 12) and it is now used, in one form or another, in a wide variety of U.K. industries.

The standard time for a job is *estimated by comparison* with a range of classified jobs, called benchmarks, whose basic times have been derived (see *Figure 10.9*) from UMS (or one of the alternative procedures). The bench marks are classified according to trade, task-area and time-range and are arranged on spread sheets as shown in *Figure 10.9.* Numerous spread sheets are necessary for a maintenance tradeforce. The job being estimated (or slotted) is given the average basic time for that range the work content of which it best fits; the times for job travel and personal needs are added. It has been shown that, over a period of time, the positive and negative job-time errors that inevitably occur cancel each other to a level acceptable for use with maintenance group incentive schemes.

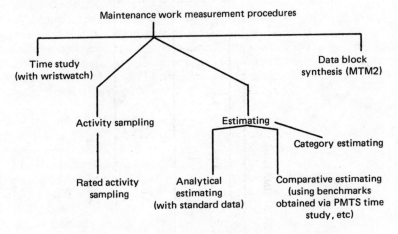

Figure 10.10. Maintenance work measurement procedures

Comparative estimating based on UMS is installed by consultants as a package (costing some tens of thousands of pounds) which includes supervision of the installation and training of the system applicators (one of whom is required for approximately every fifteen tradesmen). The objective is the raising of productivity via improved planning and tradeforce performance; a group incentive scheme is usually operated giving 30% bonus for a standard performance.

Category estimating[4] is similar to the above. The technique is based on the observation that, in workshops (or work groups) of a similar nature, the average time per job (derived from samples of about 500 jobs) remains more or less constant. Statistical analysis also reveals that the probability distribution of the job times tends to be of the logarithmic normal form. If such a distribution is measured and analysed a number of time intervals (categories or slots), which conveniently embrace the range of job times, can be assigned. *Table 10.2* shows such a time spread with the average job time calculated for each category. These job times can be standardised by carrying out a rated job-sampling procedure (see Section 10.3) for each category.

In category estimating the foreman of each working group is trained to categorise each job. The estimation is based on experience, as opposed to comparative estimating where it is based partly on experience and

Table 10.2. CATEGORY ESTIMATING

Job category	A	B	C	D	E	F	G	
Category boundary	0	0.3	0.6	1.2	2.4	4.8	9.6	19.2	
Category mid-value, h	.23	.48	.91	1.5	3.33	6.41	12.34	Obtained by sampling and statistically analysing job times
Category standard hour value	.21	.44	.84	1.61	3.07	5.91	11.38	Obtained by multiplying mid-time value by overall rating and rest factor

partly on direct comparison with similar jobs. The accuracy of the estimation procedure can be checked by periodically analysing the job times to see if they conform to a logarithmic normal probability distribution. Category estimating has been used in incentive schemes in a number of U.K. industries. It has the advantage over comparative estimating of not requiring applicators and, since standard data are not required, of being less expensive to install.

Data block synthesis[3] was developed by the MTM Association and via a pilot scheme at Ryland Bros., Warrington, U.K. It relies on the classification of maintenance work into a relatively few characteristic motions, the times for each of these motions must be determined, using MTM2, for each group of workers. It is suggested that mechanical repair work can be classified into 13 blocks of data, the first two of which are shown in *Table 10.3*. Using such blocks an applicator can quickly assess the time of a proposed job.

Table 10.3. CLASSIFICATION OF MECHANICAL MAINTENANCE WORK
INTO DATA-BLOCKS

DESCRIPTION	CODE
DATA BLOCKS	
Threaded fastener	TF
Non-threaded fastener	NTF
Handle - fit fingers	FF
" fit one hand	F1H
" fit two hands	F3H
" fit assisted	FA
" fit lifting gear	FLG
" remove fingers	RF
" remove one hand	R1H
" remove two hands	R2H
" remove lifting gear	RLG
" captive	HC
" preparation	PREP

DEFINITION OF A DATA BLOCK: Threaded fastener – TF. A single unit – such as a bolt – or a composite unit – such as a nut, bolt and washer – which joins or is joined to other items by means of mating thread. The data-block time standards are average time values for those characteristic motions obtained for each plant by direct observation, analysis, and then use of MTM-2 data tables.

Analytical estimating[5] is similar to the above but is more laborious since the time content of each job has to be synthesised from standard data stores in a card system. A low ratio of tradeforce to applicators is required.

For the repetitive parts of maintenance work, e.g. lubrication routines, the more long standing methods of time study are appropriate.

10.2.4 The Costs and Consequences of Maintenance Work Measurement

Work measurement procedures cannot be used effectively unless a great deal of attention has been given to general organisation and communication, documentation, stores organisation, job equipment, and transportation. Attention to these areas, when the maintenance department is in an initial state of non-management (which is not unusual), can greatly increase departmental productivity. This having been achieved, it does not follow automatically that work measurement should be introduced. The possible benefits (improved tradeforce performance through associated incentive schemes, improved work planning and methods, improved control) should be carefully assessed and weighed against the associated costs (of consultants, work measurement data, training of applicators and supervisors, incentive scheme bonuses, extra planning staff, etc.) and other consequences, long and short-term.

Of the maintenance work measurement procedures outlined, comparative estimating, using standard data to provide benchmarks, is probably the technique best suited to the large maintenance tradeforce. Installation and operation costs are likely to be too great for the smaller maintenance department. In this latter case, however, managerial problems are less and with well organised first-line supervision there is probably less benefit to be gained from work measurement based incentive schemes. If, nevertheless, such a scheme is being introduced (for instance, as a result of pressure from the tradeforce) then category estimating, with its lower installation and operation costs, would probably be more appropriate. This view is supported by the authors' observation that all the work measurement based incentive schemes that they investigated were effective in motivating the tradeforce for only one or two years.

Work measurement is only one of several techniques, which can be used singly or in conjunction, for increasing productivity. Such techniques must be 'bought in' since increased productivity will mean either a smaller tradeforce requirement for the same maintenance load, or an increased load for the same trade force. The success of change of this kind depends to a great extent on the political climate, both externally and within the plant concerned. Work measurement cannot be operated successfully without the full co-operation of the men involved and their trade unions.

Maintenance work measurement can be an aid to good maintenance management; it is not the panacea for maintenance ills that some practitioners in this field would have us believe.

10.3 Work Sampling in Maintenance

This is a means of obtaining information about activities or delays. The technique is based on the same statistical principles as quality control.

Snap observations of man or machine are made at perfectly random times throughout the working period studied. Thus if N random observations are made on a maintenance fitter during a representative study period, and he is found to be inactive on x of these occasions then the proportion of time for which he is estimated to be inactive is simply x/N. The precision of the estimation increases with N. Continuous observation would give a precise analysis of the time spent on different activities, but would normally be prohibitively expensive or inconvenient (and misleading, due to the biassing effect of the ever-present observer!) Conversely, a very few observations would provide information of insufficient precision. Therefore, it is important to know the value of N that will result in the desired precision. This can be obtained from the expression

$$N = \frac{4P(100 - P)}{L^2}$$

where L is the desired percentage precision and P is the *estimated* percentage of time spent on the particular activity. If P is quite unknown a small pilot study is required to obtain an initial crude estimate. As the main study proceeds the value of P, and hence of N, can be continually updated.

A list obtained by a pilot study of the activities and delays of a maintenance tradeforce is shown in *Table 10.4*. N was obtained as discussed. Snap observation tours were carried out at random intervals over a representative period of time. The results of this sampling exercise are shown in column 2 and the percentage utilisation in column 3.

Table 10.4. RATED ACTIVITY SAMPLING EXERCISE

Activity	No. of samples	Per cent utilisation	Average rating
Working with tools or equipment	205	23.6	94.8
Diagnosing/pressure testing/welding	101	11.6	100.0
Housekeeping and cleaning			
Lubricating	1	0.9	90.0
Service tours and visual inspection	2	0.2	100.0
Job consultation time (corrected)	28	3.2	75.0
Walking loaded	101	11.6	85.5
Walking unloaded	104	11.9	85.3
Resolving instructions	24	2.8	75.0
Setting up	33	3.7	94.4
Checking availability and safety			
Clerical work on job cards	4	0.5	100.0
Searching for equipment	13	1.5	89.2
Salvaging			
Loading and unloading			
Reading and checking drawings	18	2.1	100.0
Official tea break and changing allowance	87	10.0	
Authorised work study Rx. allowance	78	9.0	
TOTAL % UTILISATION OF GROUP	799	91.8	91.0
Excess relaxation allowance taken	57	6.7	
Waiting for equipment or materials	3	0.3	
Waiting because of adverse weather			
Waiting for prod. clearance			
Waiting for other tradesmen or mates	8	0.9	
Other waiting time	3	0.3	
TOTALS:	870	100.0	

$$\text{OVERALL EFFECTIVENESS INDEX} = \frac{91.8 \times 91.0}{100} = 83.5\%$$

The information thus obtained for management control was

(a) total utilisation of the group,

(b) percentage of time spent on different tasks and hence areas where improvement was needed (e.g. areas of excessive relaxation, waiting, and walking time).

If the individual work observations are also instantaneously rated (an extension of the technique, called *rated activity sampling*) it is also possible to obtain an overall effectiveness index. A number of organisations have used this OEI as a basis for group maintenance incentive schemes.

10.4 Logical Fault Finding

Repair time includes location-detection time as well as rectification time. In many fault situations, especially those where there might be a substantial chain of consequent effects, the diagnosis time can be the largest component.

The fundamental philosophy[6] of logical fault finding is to begin with the most general and to proceed to the most particular. From this emerges the following six-stage procedure.

1. Symptom analysis (determination and analysis of the effects of the fault).

2. Equipment inspection.

3. Location of faulty stage.

4. Location and removal of fault.

5. Repairing or replacing.

6. Performance testing and forwarding information for cause diagnosis.

If this is to be effectively implemented the diagnostician must be provided with adequate *training, documentation* and *test facilities.*

Input   Output

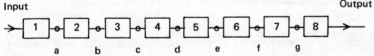

a   b   c   d   e   f   g

Test
point

Simple series circuit with input correct, output incorrect. If each check takes the same time then the checking time is minimised if each check is made at a point such that an equal number of components is eliminated whatever the result of the check, i.e. first check is at test point d.

Figure 10.11. The half-split rule

The logical identification of faults should be introduced at an early stage in the *training* of the maintenance engineer/technician. The six-stage approach should form the basis of the methods taught, which should include such fundamental techniques as the use of the half-split rule (*Figure 10.11*) and the various forms of diagnostic documentation. Practical training on complex production plant should also be available and should enable the technician to acquire the ability (a) to recognise any significant difference between the actual and ideal state of equipment and (b) to act on such recognition.

Diagnostic documentation is notable for its absence from the manuals of equipment manufacturers. It is not to be confused with conventional technical information provided primarily to facilitate corrective maintenance. Diagnosis is a separate operation from repair and it demands a specific type of documentation. This can make use of a number of formats (e.g. algorithms, decision diagrams, block diagrams) constituting a diagnostic procedural package which, to be effective, must subscribe to the fundamental philosophy of proceeding from the most general to the most particular. It is also essential that the diagnostic formats

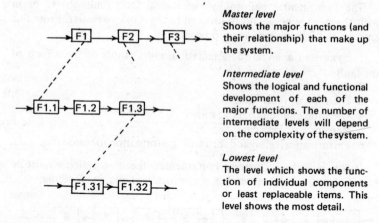

Master level
Shows the major functions (and their relationship) that make up the system.

Intermediate level
Shows the logical and functional development of each of the major functions. The number of intermediate levels will depend on the complexity of the system.

Lowest level
The level which shows the function of individual components or least replaceable items. This level shows the most detail.

Figure 10.12. An equipment identification system for diagnostic documentation

should reflect the functional characteristics of the equipment. A sophisticated system (*functionally identified maintenance system*[7]) based on the above principles has been developed by the Royal Navy; the identification system for FIMS is shown in *Figure 10.12* (see also *Figure 3.2*). A number of different diagnostic formats (some of which are reviewed below) are used at the different levels of this system.

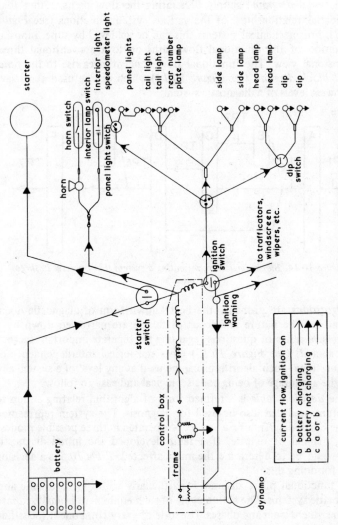

Figure 10.13. Functional layout of car electrical system (part diagram)

starter

horn switch

interior lamp switch

interior light

speedometer light

panel lights

tail light

tail light

rear number plate lamp

side lamp

side lamp

head lamp

head lamp

dip

dip

horn

panel light switch

dip switch

ignition switch

to trafficators, windscreen wipers, etc.

starter switch

ignition warning

control box

frame

dynamo

battery

current flow, ignition on

a battery charging

b battery not charging

c a or b

With relatively uncomplicated equipment one of the conventional engineering representations, without any amendment may suffice for diagnostic purposes. Complex plant may be more readily analysed by a *functional diagram*[7] simply illustrating the flow paths, rather than the spatial relationships, of the various system functions (see *Figure 10.13*). For mechanical systems this can be achieved by superimposing illustration of the functional flow paths on to a conventional three-dimensional view. The functional flow concept gives rise to the *functional block diagram* (see *Figure 10.14*) which can be used at all save the lowest levels of a diagnostic system.

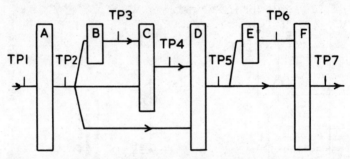

Figure 10.14. *Block diagram representing a system that operates in three modes*

Algorithms are probably the best known form of diagnostic documentation. The nature and locus of failure are narrowed down by a progressive series of questions regarding the separate important factors. A flow chart (see *Figure 10.15*) plots the logical interdependence of the questions. Such algorithms can be used at any level of a system and have the advantage of being concise, logical and easy to follow.

The *decision table* is a refined form of algorithm relating action to symptom, and can also be used for diagnosis. The system represented by the block diagram of *Figure 10.14* operated in three possible modes: aural, manual or remote. If a fault developed the initial diagnostic step would be to determine the mode affected. *Table 10.5* is a decision table for doing this.

A functional phase diagram is particularly useful for hydraulic and pneumatic systems. The action cycle of such equipment is analysed into its constituent steps and phases. The state of every component (classified as a working, controlling, or other part) is plotted for each phase, against time. In the event of a fault the diagram is used for establishing the point at which the cycle of action was interrupted. This idea is best explained by a simple example.

217

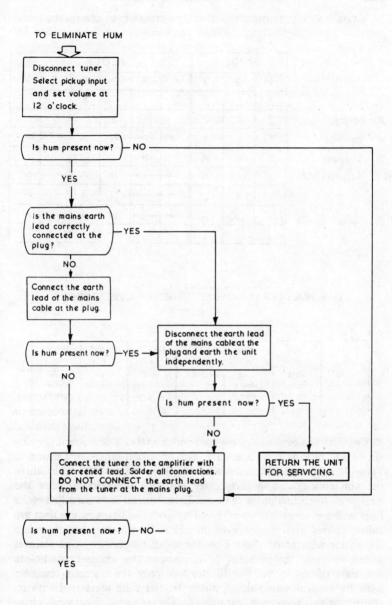

TO ELIMINATE HUM

Disconnect tuner
Select pickup input
and set volume at
12 o'clock.

Is hum present now? — NO

YES

Is the mains earth
lead correctly
connected at the
plug? — YES

NO

Connect the earth
lead of the mains
cable at the plug.

Is hum present now? — YES →

NO

Disconnect the earth lead
of the mains cable at the
plug and earth the unit
independently.

Is hum present now? — YES

NO

Connect the tuner to the amplifier with
a screened lead. Solder all connections.
DO NOT CONNECT the earth lead
from the tuner at the mains plug.

RETURN THE UNIT
FOR SERVICING.

Is hum present now? — NO —

YES

Figure 10.15. The use of an algorithm in logical fault finding

Table 10.5. DECISION SCHEME FOR SYSTEM REPRESENTED IN FIGURE 10.14

MODE 1	2	3	BLOCK A	B	C	D	E	F
X	X	X	S			S		
X	X	✓			S			
X	✓	X	← —		N	A	— →	
X	✓	✓		S			S	
✓	X	X	← —		N	A	— →	
✓	X	✓	← —		N	A	— →	
✓	✓	X	← —		N	A	— →	
✓	✓	✓						

X   Faulty
✓   Correct
S   Suspect
N A   Inadmissible

Table 10.6. OPERATIONAL SEQUENCE OF A PRESS TOOL

1 2	3 4	5 6	7 8	9 10	11 12	13 2 3
PHASE a.	PHASE b.	PHASE c.	PHASE d	PHASE e	PHASE f	PHASE a.
work put in	work locked	down stroke slow	rapid return	work unlocked	work removed	new cycle starts

Consider the case of a press tool which makes a downward pressing stroke and returns at a faster rate. The operation sequence is shown in *Table 10.6*. The vertical lines represent those parts of the cycle where the non-work-executing parts carry out their functions to allow the next work-executing phase to occur. These vertical lines are called phase lines and are given odd numbers. The spaces in between the lines are called phases and receive even numbers. In this instance there are six phases with seven phase lines for actual functioning. One phase is added to show the positions of all components when at rest before the start of the cycle. For illustration, only the clamping cylinder with its control valve, and a piston to carry the press tool with its control valve, have been considered. The resulting functional phase diagram is shown in *Figure 10.16*. Note, similar duty parts have been grouped together, i.e. work-performing parts, phase-changing parts.

The diagnostic procedural package selected will depend on the expected diagnosis time and the unavailability cost. If both factors are significant then the application of an advanced technique (e.g. the functionally identified maintenance system) might give a large return on the investment in the necessary documentation and training of

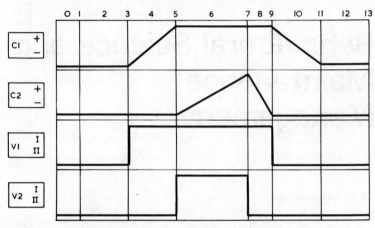

CI is clamping cylinder controlled by valve VI

C2 is pressing cylinder controlled by valve V2

Figure 10.16. Functional phase diagram

staff. If the anticipated diagnosis time is smaller, less complex techniques (functional diagrams, algorithms, etc.) can be of great value. What must be emphasised is that both the diagnostic documentation and the *test facilities* should be prepared as an integral part of the plant design stage rather than as an afterthought.

REFERENCES

1. BS 3138, *Glossary of Terms in Work Study,* BSI (1969)
2. Stewart, H. V. M., *Guide to Efficient Maintenance Management,* Business Publications (1963)
3. Rendel, G. D., 'MTM-2 in Maintenance Engineering', *Iron and Steel Institute Publication 118* (1969)
4. McKinnon, R., 'Maintenance Repair', *Business Administration,* 44–47, Apr. (1971)
5. Kelly, A., 'Work Measurement for Maintenance Manager', *Factory,* Sept. (1972)
6. Bayliss, R. G. and Langham-Brown, J. B., 'Prevention Plus Fault Diagnosis Equals Availability', *6th National Maintenance Conference,* Nov. (1972)
7. Langham-Brown, J. B., 'Diagnostic Documentation is Vital to Fault Diagnosis', *Factory,* Nov. (1972)

Chapter 11

Behavioural Science and Maintenance Management

11.1 Introduction

The most important factors governing the effectiveness of a maintenance organisation are the diligence and skill of the tradeforce. Sound planning and engineering are essential (that is why so much of this book has been devoted to organisation theory, work planning and industrial engineering techniques) but equal attention must be given to analysing the needs of the individual worker and to creating a work situation which satisfies those needs. Neglect of this latter area has been the cause of much industrial unrest.

The difficulty is that the needs of the individual and the needs of the organisation often conflict. On the one hand classical organisation theory suggests that work should be divided into simple units, each of only a few motions, thus facilitating specialisation, the worker being motivated by monetary incentives. On the other hand behavioural science suggests[1] that such arrangements are inimical to psychological health. A number of studies have aimed at identifying those factors in the industrial situation which are conducive to the generation of a contented and motivated workforce.

Maslow[2] identifies and arranges in order what he considers to be the needs of the individual, i.e.

Higher needs	5. Self fulfilment
	4. Autonomy
	3. Self-esteem

Basic needs   2. Sociality

1. Security

Maslow argues that when the first need, security of employment and sufficiency of income, is satisfied the individual's attention turns to social needs, i.e. group membership and acceptance. When these also have been satisfied the next step is for self-esteem, having some attribute, status or skill which others are believed to recognise. Next comes the desire for autonomy, for a greater opportunity of being one's own boss on the job, and for a corresponding reduction in the constraints upon individual freedom imposed by an immediate supervisor. Finally, and hardly ever fully achieved, comes self fulfilment — the maximum development, at work, of all the individual's skills, abilities and attributes. Although Maslow's approach is over-simplified, it does provide a useful framework for assessing an industrial situation as it affects the individual.

Herzberg[3] also divides the needs of the individual into basic (biological) and higher (growth) needs; he furthermore identifies and

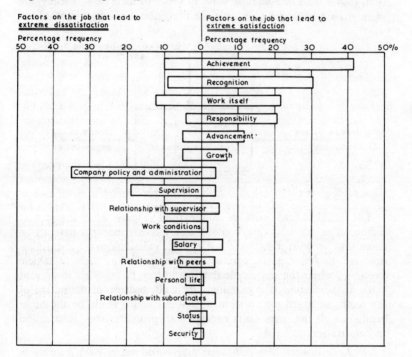

Figure 11.1. Factors affecting job attitudes[3]

quantifies the factors affecting those needs (see *Figure 11.1*). He emphasises that it is factors associated with the *higher* needs that can affect job *satisfaction* and that, in the industrial setting, these factors are to be found in the job content. Factors which are associated with the *basic* needs affect job *dissatisfaction* and these are to be found in the job environment. Herzberg believes that while different factors, under different circumstances, can be used to 'get work done' only the factors associated with job content are the true motivating ones. He argues that this is because a 'motivated worker is responding to an internal stimulus, a desire to carry out the work'. He points out that if a man works as a result of an external stimulus he will tailor the resultant work in a way that suits him best (he minimises the pain of working), and his preoccupation at work is with the balance of the external stimulus and his effort rather than with the organisational need. The man 'gets work done' but the motivation is with someone else. This motivation-hygiene theory provides an insight into the conflict of the organisation and the individual. The theory suggests that, rather than *rationalising* work in order to increase efficiency, the work should be *enriched* in order to bring about effective motivation.

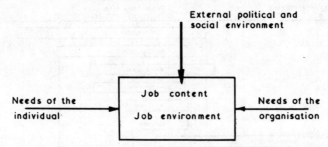

Figure 11.2. Meeting point of the individual and organisational needs

The problem can now be summarised. *Figure 11.2* shows that factors affecting both job content and job environment have to be taken into account if both the needs of the individual and the organisation are to be met. In certain areas of industry, such as vehicle assembly, while the environmental factors can be adjusted to prevent work dissatisfaction it is extremely difficult, because of the nature of the work, to inject job enrichment*. Maintenance work, on the other hand, has all the ingredients necessary to promote work satisfaction and motivation.

* Several experiments[4] have been successfully carried out on 'group assembly' in Scandinavia (see also Section 11.2).

It is the authors' observation that in many maintenance situations management have over-emphasised the needs of the organisation and have thus *downgraded* the job content. Attempts have been made to motivate the work force by monetary incentive (based, for example, on UMS). It is not surprising that although productivity rose considerably in the short term (because of improved organisation rather than improved performance) it fell back again in the longer term. Management were then left with improved control but no *true* motivation and the search for ways of improving productivity continued.

If job-enrichment is the way forward then it has been suggested[4] that a manager should:

1. Replace detailed instruction by clarification of objectives.
2. Increase responsibility and provide greater chance of achievement by making the jobs of planning, organising, directing and controlling a joint function with employees.
3. Study the organisation of jobs and try to design them so as to give greater satisfaction of human needs.
4. Replace control activities, by those which seek to emphasise the manager as a helper/supporter/tutor, in order to develop abilities.
5. Set out to build effective teams in his workforce.

To operate such a flexible system a great deal of tradeforce self-discipline is required. However, this 'right attitude of mind' is formed by the external political and social climate as well as the job content and job environment and *if such external factors are reactionary then no amount of job-enrichment will promote enhanced productivity.*

11.2 Swedish Views on Job Reform

Over the last decade the industrial experience of Sweden has produced some very interesting ideas. Managers and researchers[4] in that country have found that while many of Herzberg's notions are valuable they are not without limitations. They argue that Herzberg has not studied the relationship between motivation and performance and that his views on payment-by-results are questionable. They find that a properly constructed payment-by-results system (a group system in the case of maintenance) is a powerful motivator. Furthermore, although Herzberg is not enthusiastic about group work or worker participation both of these methods are, in Sweden, widely practiced with good results. They conclude that, since Herzberg's approach specifically excludes much of what seems to be most effective in Sweden, his theories are not derived from practical reality.

Although this book is concerned with the maintenance function, organisational changes and job re-design in the production function in Sweden merit brief description because of the relationship between the two functions. The aim of Swedish unions and management has been to improve the lot of the shop floor worker by implementing points 1–5 of the previous section. This has meant a *complete re-think* about methods of work and about the organisation of such work.

In summary, job enrichment has been obtained by

1. Extending the work cycle with additional production tasks.
2. Integrating production and auxiliary tasks such as simple short-cycle maintenance tasks.
3. Decentralising authority and responsibility so that the worker has more control over his own actions.

The Swedes have found that extending the work cycle (say, from 2 minutes to 13 minutes and above) does not, for most people, lead to loss of efficiency. Integration has been favourably received by the production worker and is particularly important in the case of functional groups such as maintenance departments. Decentralisation of authority has probably had the most favourable overall impact since, in addition to enriching the work, it has also resulted in shop-floor workers having a more responsible attitude towards company objectives, management and equipment. This becomes particularly significant from the point of view of maintenance when it is realised that a large proportion of the corrective work load arises from maloperation.

The above techniques are encapsulated in the Norwegian and Swedish concept of autonomous groups. This is based on 'job rotation' and was first used in the production industries where job enlargement by other methods was difficult. The concept has now been adapted to other types of production work. The autonomous group method differs from the normal methods of job rotation in that job rotation occurs as the work situation demands it. The group has considerable autonomy of action and appoints its own team leader; changes within the group and outside it are conducted via informal *participation* procedures. It has been found that group work is most successful when it is arranged along the production line (with the grain) rather than across the production line (across the grain).

Such fundamental changes in the work situation have compelled changes in the supervisory role and in organisation and planning methods. Although the supervisor is still necessary his role has changed; he has had to learn to discuss, not order. Increasingly, he presents and discusses problems with the groups rather than offering ready-made solutions. One of his main functions is co-ordination between the groups and the back-up services and planning systems. Careful study was also made of the planning system. This concluded that overplanning and control dehumanised the work and was also ineffective. In a number of companies the organisation was 'resimplified', the central idea being that the less important administrative tasks were pushed down the organisation and were dealt with at the appropriate level. Administrative resources were then freed to concentrate on important strategic and economic factors.

Clearly, the general advantages of 're-simplification' of the organisation, of participation, and of decentralisation of authority, apply to both the maintenance and production functions. Some of these changes, however, affect maintenance work more directly. Integration of the simpler, routine, maintenance tasks with the production work not only enlarges the production job, but also fosters the production worker's commitment to the plant. The maintenance worker's particular skills are utilised more efficiently. For practical reasons, such as the complexity of the plant, there are limits to the process of integration and the need for large maintenance departments still remains.

Integration within maintenance departments has taken two principal forms.

1. Considerable effort has been devoted to reducing inter-trade demarcation, e.g. development of the 'electro-mechanic' who, as well as carrying out the electrician's work, undertakes the simpler fitting tasks.

2. In the more substantial process companies, operating a number of large self-contained plants, efforts were made to form autonomous, multi-trade-supervised, maintenance groups (see *Figure 5.6*). It was found that while such arrangements facilitated co-ordination of multi-trade maintenance work they led to isolation of specialised tradesmen and poor labour utilisation in the more specialised work. A number of companies have overcome this problem by adopting the so-called 'matrix organisation'. Every individual is attached to a certain maintenance plant-group and also to a certain occupational group. This organisational technique has now been in operation for a number of years and has given good results.

Although there is much to be learned, with possible practical benefit, from the Swedish experience it cannot be emphasised too strongly that their industrial climate is very different from our own. Workers, trade unions, and management recognise the necessity to co-operate in the achievement of jointly agreed objectives directed towards efficiency and profitability — perhaps this is where we have the most to learn from Swedish industry.

REFERENCES

1. Argyris, C., *Personality and Organisation. The Conflict between System and the Individual*, Harper and Brothers, New York (1957)
2. Maslow, A. A., *Motivation and Personality*, Harper and Brothers, New York (1954)
3. Herzberg, F., 'One more time: How do you motivate employees?' *Harvard Business Review*, Jan./Feb. (1968)
4. Swedish Employers' Confederation (Technical Department), *Job Reform in Sweden* (1975)
5. Johnston, A. V., *Motivation of Labour, Staff and Management*, The Iron and Steel Institute, Publication 118 (1969)

Chapter 12

Case Study 2. Motivation and Reorganisation of a Maintenance Trade Force

12.1 Introduction

This case study described the application, in 1964, of a Maintenance Incentive and Planning scheme in one of the maintenance departments of the BICC parent company.

In order that the full impact of this system can be appreciated it is first necessary to outline the structure of the BICC group. This is shown in *Figure 12.1* where it can be seen that the organisation consists of four main sections containing a number of separate self-contained units of varying size and complexity. Although the main activity of the group is the manufacture of electrical cables of many different types, there are also many group companies engaged in a wide range of other activities from constructing road tunnels to making rubber bands.

12.2 Background of Unit

The particular unit from which this study is taken produces household and general cables and employs approximately 2500 people on a 26 acre site. A maintenance engineering work force of 116 was eroded to 91 by reduction of contract labour and natural wastage after the installation was completed.

* Contributed by R. Warburton, Industrial Engineering Manager, British Insulated Callender's Cables, Ltd., Prescot, Lancs.

227

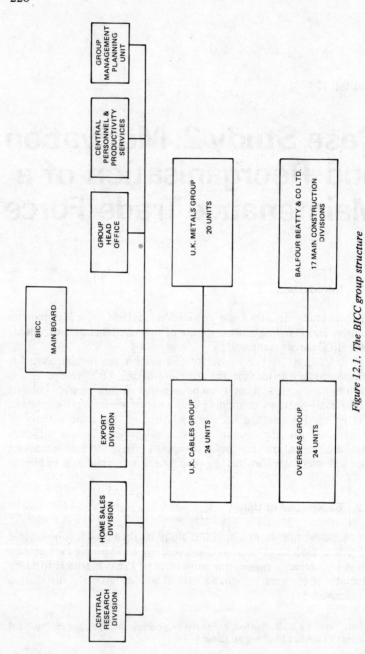

Figure 12.1. The BICC group structure

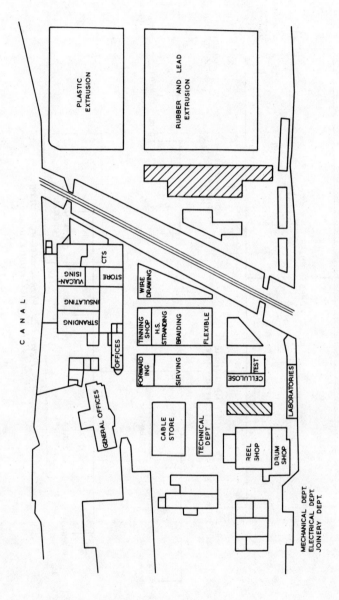

Figure 12.2. Plan of works. (Scale 1 in = 240 ft approx.)

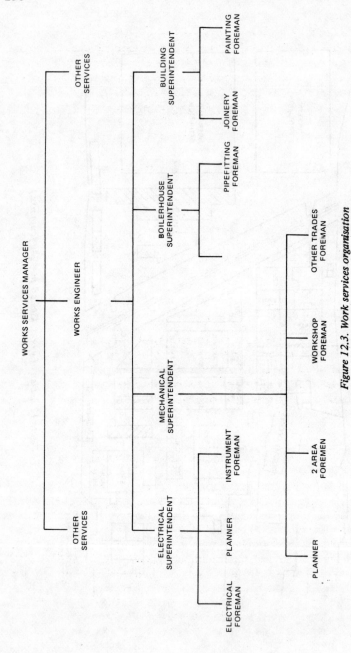

Figure 12.3. Work services organisation

A fixed bonus payment of approximately 14% of base rate was paid to maintenance engineering personnel without any measure of achieved performance (a situation which, at that time, was typical of the whole BICC group). However, in the case of production workers, the company policy, which was well established, was to provide a bonus opportunity of one-third of basic rate for the achievement of standard performance (100 BSI).

The layout of the unit concerned can be seen in *Figure 12.2*, with the location of the workshops shown as shaded areas. The works services management and supervisory organisation was as shown in *Figure 12.3*.

12.3 Survey

The inequity of bonus opportunity, which existed between the engineering maintenance and the production workers, led during 1963 to a request from the engineers to increase the existing fixed bonus payment. A forward looking unit management then negotiated acceptance of some form of measurement in return for an increase in bonus payment.

In order to establish the potential for improvement a survey was undertaken during which the main technique used was rated activity sampling. This indicated that the departmental performance level was extremely low (approximately 45 BSI) and it was concluded that the introduction of some type of incentive scheme based upon work measurement should raise the efficiency of the department very considerably and that either a reduction in manning of 25 people or an equivalent increase in output could be achieved.

12.4 Setting up the Installation

At this point the reasons for desiring closer control of maintenance work were identified as the following:

1. Increased investment in new plant added to the maintenance load and demanded an improvement in maintenance efficiency.
2. Production stoppages were costly.
3. Maintenance payment arrangements had to be brought into line with those of production workers.

At the time, maintenance schemes were virtually non-existent in the U.K. and in order to gain knowledge of the types of schemes

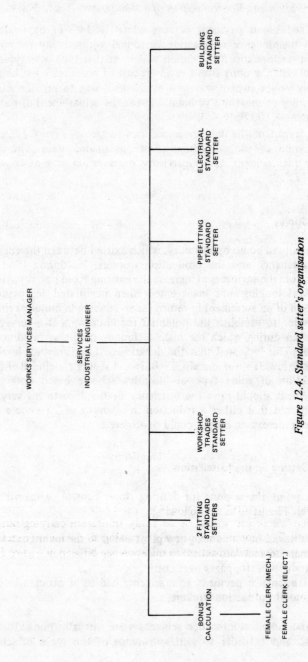

Figure 12.4. Standard setter's organisation

already in operation it was decided that a team led by the Works Services Manager would visit Sweden, where there were already many schemes in operation. Following this visit negotiations with the union representatives were resumed and in May 1964 a consultant with a knowledge of a suitable measurement system was engaged.

Nine months was the estimated time needed to complete the installation process, which encompassed the following tasks:

1. Training of the tradesmen who were to carry out the measurement and eventually become responsible for setting the target time values (these men were transferred from the hourly paid to the salaried staff).

2. Collecting all the necessary studies of typical jobs for all the trades to be covered by the scheme (approximately thirteen hundred studies in total).

3. Operating a dummy run of the scheme to ensure that sufficient measurement had been done.

4. Complete overhauling of the organisational aspects of the departmental operation.

In order to be fair to both unions and management, it was agreed that the first three months of the scheme's operation would be treated as a trial period. There could have been a termination by either side at the end of that time.

It was decided that bonus would be paid on a group basis in order that accurate performance figures could be obtained without extending the bonus pay period. The pay groups were established as follows:

Group 1	—	Fitters
Group 2	—	⎧ Sheetmetal workers ⎨ Welders ⎩ Blacksmith
Group 3	—	Pipefitters
Group 4	—	Electricians
Group 5	—	Joiners

The final preparatory requirement was the selection and training of tradesmen whose job would be to study tradesmen carrying out jobs and eventually to compile a wide-ranging catalogue of standard times. The structure of the standard-setter organisation can be seen in *Figure 12.4*.

12.5 Content of the Installation

Maintenance work does not lend itself to the use of conventional work study techniques. It was decided therefore that an integrated programme had to be established in which the time standards and payment systems were based upon a solid organisational foundation. This programme had two main aspects, (a) organisation and (b) work measurement.

12.5.1 Organisational Aspects

These were considered as important as work measurement in raising the efficiency of the department.

Stores It is unrealistic to expect men to work under incentive conditions without an efficient stores organisation. The geography, manning, stock records, and special tools and spares aspects of the stores organisation were all carefully examined.

Tools The equipment which a tradesman has to use should be in good condition and suited to the job requirements. A standard tool kit was purchased for each tradesman, together with a suitable means of storage and transport (in this case, a metal tool box and trolley).

Shop layout The facilities within the workshop were reviewed to ensure that there was sufficient machinery and storage space. Also that there was space for work on large components and that work bench facilities were adequate.

Chain of command Responsibility for areas of work were clearly defined, e.g. 'who should disconnect the electric motor', thus avoiding confusion and delays on the job. Also, the foremen's responsibilities were clearly defined in writing, thus ensuring that they were fully aware of their duties in regard to the operation of the department.

Deployment of labour During the survey which preceded the installation it was established that more economical use could be made of labour by setting up centralised trade workshops as a base from which all tradesmen would operate. Not unexpectedly this created some problems within production supervision who previously had received personal service from resident tradesmen. However, it was quickly appreciated that the overall service to production was improved by the combination of centralised operation and the application of a properly measured and planned maintenance operation.

Requisitioning procedures   A procedure was established for transmitting all breakdown requests by telephone. All other requests, for installation or overhaul work, were made on a formal requisition which was initiated in the production department concerned and which was routed via the Works Engineer's department.

M.O.W. WORK ORDER						35010	
Requested by	Date Rec	Location	M/c No.	Dept	Date	Date
Authorised by	Date needed	Cost Ref	No. of Men	Trade	Time	Time
					M/c stop	M/c ready
Fault or Work Required						
Planned Work						
Changes to planned work					Hrs allowed	
					Hrs used	

Figure 12.5. Maintenance work order

Planning In order to ensure that major installation jobs were properly organised a planning procedure was established. This was a fairly detailed system in which mechanical and electrical activities were scheduled, using the measurement system to effect accurate prediction of activity completion.

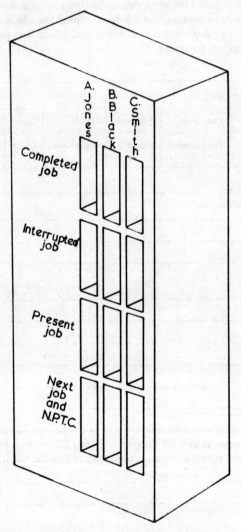

Figure 12.6. Work order board

Work order procedure It was essential to put the issue of work on to a formal basis such that an adequate record, of the work content and allowed time for the job, was made. The system which was set up also enabled performance, measured work percentage and cost control data, to be accumulated. (*Figure 12.5* shows the front of a typical work order.) The start and finish time of the job could be recorded by the tradesman, on the reverse side. Also, a *non-productive time card* (NPTC) was designed on to which all waiting time was booked, in coded form, by the tradesman, thus facilitating analysis of different types of waiting time. These work order cards were issued via a board containing a series of pigeon holes, one set for each man, as shown in *Figure 12.6.* This board classified the cards into 'next

Figure 12.7. Typical office layout

job', 'present job', 'interrupted jobs' and 'completed jobs', thus providing the foreman with a complete visual aid as to the state of the work load in his department at any given time. *Figure 12.7* shows a typical office layout with the foreman, the standard setter and the work order board situated adjacent to each other.

12.5.2 Work Measurement Aspects

The work measurement system was based upon *methods time measurement* (MTM, see Section 10.2). The aim was to provide a catalogue of

BICC   CATALOGUE   CODE TRADE FITTING

JOB TYPE   GEAR BOXES COUPLING AND MOTORS

(0.00) HRS 0.1 (0.20)	(0.20) HRS 0.3 (0.40)	(0.40) HRS 0.5 (0.60)	(0.60) HRS 0.7 (0.80)
F.9 Estler Buncher Renew flexible coupling	F.157 Remove Carter gearbox using step ladder 2 men	F.85 Wire drawing rod machine: Reposition gearbox coupling using hammer. Drift and hydraulic rack. Remove guard and re-assemble. 2 men	F.49 Nashoba striping machine. Change Carter gearbox. Transfer conversion equipment pipes. Transfer speed indicator clock & reset. 2 men
F.143 Carter Gearbox Type F.12 VS change control speed handle from L. hand to R. hand or reverse	F.50 Pirelli lead Press remove 1 variable gearbox	F.91 Robertson medium wire machine. Remove guard and cover plate. Slide selector into next gear ratio and re-assemble. 1 man	F.130 Pirelli lead press: Remove 2 variable speed gearboxes: Remove 2 keys and couplings. 2 men
F.156 Remove Carter gearbox from gearbox bed 2 men	Remove 6-½" coupling bolts.		F.223 Iddon Extdr align & re-shim motor 2 men
F.178 half coupling Outside dia. 2½" bore fit coupling to shaft 1 man	Remove 4-." holding down bolts and dowel pins. Sling gearbox and lower to floor. 2 men	F.207 Carter gearbox — Remove servo motor assembly. Replace and test run for 4 min. 1 man	F.224 Iddon Extdr. Replace motor on baseplate. Align couplings and secure. 2 men
F.181 half coupling Fit wood block in bore of coupling mark out centre punch 4 equal spaced holes 1 man			F.225 Iddon Extdr. Draw-off half coupling from gearbox with pickervant three arm withdrawer. 2 men
F.183 5" dia coupling set up ¾" setscrew and steel block with packing, withdraw key, then withdraw coupling.			

Figure 12.8 Spread sheet (catalogue)

standard times for typical maintenance jobs. *Figure 12.8* shows one page of a typical catalogue and it can be seen that the job descriptions are classified in two ways, task type and time group. By means of this classification the standard-setter can set standards for jobs by making appropriate comparisons with jobs in his catalogue and selecting the time group for the work content of the job in question.

The time values contained in the catalogue contain work content at the job site only and in setting a job time for issue to a tradesman it is necessary to add time for preparing tools and equipment, for walking to and from the job site and for the normal relaxation allowances associated with an incentive scheme. A typical build up of a standard time can be seen in *Figure 12.9*. Note that the final time for the job is again put into a time group as in the catalogue. This is done firstly to

Work Description	Hours
Medium wire machine, Gear ratio change (F.91)	0.5
Travel time	0.1
Preparation time	0.05
	0.65
Relaxation allowance, 15%	0.10
Total time	0.75
Time group time	0.7

Figure 12.9. Typical standard time build-up

enable comparisons to be made in the catalogue and secondly because the aim is not to set exact standards for every job, but rather to assign jobs to a time range such that over a reasonable pay period, and for a group of men, they will be accurately representative.

The system of measurement was successfully applied to a wide variety of maintenance tasks e.g. corrective and emergency maintenance, installation. It was particularly important to have an accurate system of measurement for preventive maintenance work in order that predictions regarding machine availability could be fulfilled, thus establishing the conditions whereby production supervision would be willing to release machines for inspection.

12.6 Results of the Installation

This was completed in February 1965 and as a result the following was achieved:

1. Departmental performance was improved to around 100 BSI.
2. Cost per standard hour was reduced by a half.
3. Net departmental labour force was reduced by 25, i.e. from 116 to 91.
4. Machine downtime was reduced,
   (a) mechanical, from 1000 h to 680 h,
   (b) electrical, from 240 h to 150 h.

The unit activity was also raised by 14%. Contract labour was no longer employed.

12.7 Extension of the System

Following the success of the installation, in the unit described in this case study, the system was extended to many other parts of the BICC group. Some fifty schemes have been installed, covering three main trade groups, fifteen trades and approximately 1100 men.

12.8 Conclusion

The benefits which accrue from the application of this type of maintenance installation are as follows:

For the worker: better conditions for performing maintenance work,
standard times for jobs before they are started,
consistent time standards,
payment based upon performance.

For the company: more effective maintenance,
better control over maintenance work,
lower maintenance costs,
reduction in machine downtime,
availability of data for analysis, to determine improved maintenance policies.

The ingredients for a successful installation of this type are:

determined management support for the system,
supervision willing to accept changes in method of operation,
competent industrial engineering effort,
a co-operative climate between management and unions.

Chapter 13

Case Study 3.
A Maintenance
Management
Investigation

13.1 Introduction

The following case study[1] concerns an investigation into the maintenance management procedures that were being used to control the
availability of a fleet of diesel equipment. This case study is being
introduced as the concluding chapter of the book because it illustrates
how many of the ideas and techniques of the earlier chapters were used
in a practical investigation. For example the use of block flow diagrams
to model the maintenance/production/availability situation (see *Figures
1.3* and *13.1*), Weibull analysis to diagnose the cause of failure and
prescribe solutions (see Section 2.5 and *Figure 13.3*), queueing theory
to model the repair situation (see Chapter 6 and *Figure 13.5*). The
understanding of these basic ideas and techniques was used to provide
a solution to a complex and expensive problem.

13.2 Background

The investigation was carried out by A. Kelly and a team of four
students from the University of Manchester during the summer vacation
of 1973. The company involved in the project was concerned with the
large scale mining of copper ore and other minerals and at the time of

242

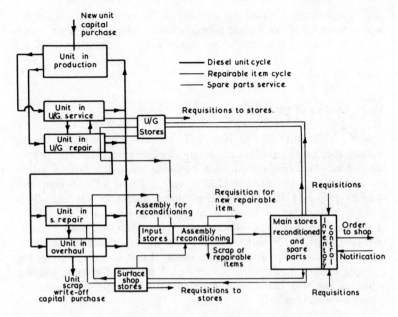

Figure 13.1. Diesel maintenance system

the investigation was expanding its copper mining operations. An important part of the expansion was the growth in the use, underground, of diesel equipment; the capital value of such equipment stood, in 1973, at £4.5M. The rapid expansion of production arising out of the use of diesel equipment had caused numerous production, maintenance and organisational difficulties which had been reflected in maintenance costs and in low availability.

Preliminary discussions with the engineering management suggested that the problem was complex and would be all the more difficult because the management procedures had evolved over a period of time and had not been charted. As a result of these discussions it was decided that the programme should take the following form:

Familiarisation with the production equipment, methods and policies.

Mapping of the work, instruction and information flow paths of the overall diesel maintenance system.

Costs and availability analysis of the diesel maintenance system in order to identify the main problem areas.

More detailed investigation where necessary.

Discussion of the findings in order to recommend improvement.

The following is an edited account of the investigation, *simplified in a number of places for the sake of clarity,* followed by a discussion which includes recommendations made to the management of the company.

13.3   Outline of the Mining Operation

It was decided that this project would concern itself mainly with the maintenance problems of the copper mining operations. The vast copper ore bodies were being mined, some half-mile below the surface, into a network of large open spaces (130 ft wide x 200 ft x 700 ft high) and support pillars (120 ft wide). The mining operation was basically a drilling, blasting and transportation process and could be considered as a batch process since blasting took place over one hour periods, once per shift, three shifts per day, six days per week. The diesel equipment (see *Table 13.1*) was an essential part of its operation, for example drilling equipment was mounted on diesel carriers and the exploded ore was hauled and dumped into the main transportation system by diesel loaders.

Table 13.1. DIESEL UNITS IN THE COPPER MINING AREA

Diesel loaders	
8 cubic yd bucket	1
5 cubic yd bucket Type A	13
5 cubic yd bucket Type B	5
5 cubic yd bucket Type C	7
4 cubic yd bucket	1
Drilling equipment	
(Multi-boom Jumbo drills)	7
Work platforms for drilling	
equipment	14
Trucks	5
Service vehicles	14
Diesel locos	4

Because of the high cost of lost production the operating policy was to provide spare diesel units to compensate for unreliability and this largely prevented lost production. The greatest problem in this respect was the diesel loaders, which operated in particularly arduous conditions, coupled with the fact that the operators were on an incentive bonus scheme. Thus the level of availability of individual production diesel loaders seemed particularly low compared to other industries. This, in conjunction with the random incidence of breakdowns and repair times, caused large fluctuations in the number of loaders available at any one time. The production policy was to use all the machines that became available for production. In order to maximise the operating flexibility, loaders were not attached to any individual operator or group of operators. Thus over a period of time a particular loader might have been used by every operator.

13.4 The Diesel Maintenance System

The accepted objective was given by management as the minimisation of the direct cost (men, spares and equipment) of diesel maintenance for a particular level of unit availability. This level of unit availability varied with the type of unit (e.g. 70% for the diesel loaders in the copper orebodies) and was assumed to be that level at which maintenance costs (i.e. the sum of the direct maintenance costs and the costs associated with unavailability) were optimised. No information was put forward by management to justify this assumption but since production had to be planned some considerable time ahead it was necessary to have realistic availability figures on which to base future unit requirements.

The maintenance policy had evolved out of the production policy and operating methods, the underground environment, and the geography of the situation. The *underground repair* policy was to carry out work in the quickest possible way, which meant in many cases the direct substitution of assemblies, sub-assemblies and components of the diesel units. Because of the difficulty and cost of transporting units between levels, area workshops (supported by sub-stores) were located near the production areas. The area workshops had limited facilities and it was necessary for *major repairs* and *overhauls* to be carried out in a well-equipped surface workshop (supported by the central stores) situated about 1 mile from the main haulage shaft. This surface-workshop had a section set aside for the *reconditioning* of the failed parts which were returned from the underground area workshops and also from the repair/overhaul section of the surface workshop.

The preventive maintenance policy was based on *servicing* and *overhauling* of the units. The servicing (lubrication, inspection, adjustment) of the diesel units was carried out in the area workshops at a frequency determined by operating hours. Repairs required as a result of service inspections were carried out immediately. Overhauls involved a complete strip down and rebuild of the diesel units, the frequency of which was determined by the operating hours and periodic cost trend.

From the foregoing it will be appreciated that the diesel maintenance system could be represented (see *Figure 13.1*) as a complex cycle of diesel units in *production* (available) or in *service, repair, major repair, overhaul* (unavailable) or *waiting* for maintenance attention (unavailable).

The reconditioning of assemblies could also be regarded as an operations cycle necessary to support the main diesel unit cycle. The reconditioning cycle is shown in *Figure 13.1*, superimposed on the unit cycle.

The storage and flow of spare parts necessary to support the maintenance activity are also shown on the same diagram.

The cycle outlined in *Figure 13.1* was useful in as much as it outlined the overall operation of the maintenance system and was therefore used as a basis for the investigation. For such an investigation to be objective it was felt necessary firstly to analyse the *direct* maintenance costs and then the equipment unavailability levels (which reflect the *indirect* maintenance cost).

13.5 Cost Analysis of the Diesel Maintenance System

The costing information used in this analysis was extracted from the maintenance budgetary and cost control system. This system was investigated but will not be discussed in detail. However it is necessary to say that the costing system had been established some years earlier and did not provide sufficient relevant data quickly enough to effectively control the diesel maintenance system.

13.5.1 Direct Cost of Diesel Maintenance

The average direct cost of maintaining the whole flow of diesel equipment (for all mining operations) per four-week period during 1972/1973 was £230 000. This amounted per year to 60% of the capital value of the diesel equipment and to 40% of the total maintenance bill for the whole mine including the vast surface operation. The major part of the total cost came from the copper mining area (*Table 13.2*) and it can be seen from *Table 13.3* that this was as a result of the diesel loaders.

Table 13.2. COST ANALYSED BY UNDERGROUND AREAS

Underground area	% units	% cost
Copper mining	47	57
13 Level	10	15
19 Level	16	10
Other areas	27	18

Table 13.3. COST ANALYSED BY MAJOR DIESEL LOADERS IN THE MINING AREA

Unit	Total cost period (× £1000)	Unit cost period (× £1000)
5 yd Type A (×13)	50	3.8
5 yd Type B (×5)	21	4.2
5 yd Type C (×7)	22.7	3.2
8 yd (×1)	2.6	2.6

Table 13.4 COSTS ANALYSED BY MAINTENANCE SECTION

Maintenance section	Men	Materials	O/Heads	Total
Servicing	7	1	2	10
Repair	20	28	4	52
S. Repair/overhaul	5	6	4	15
Reconditioning	6	12	5	23
Total	38	47	15	100

It was also felt necessary to establish the distribution of the total costs around the maintenance cycle and to classify such costs into the constituent parts. This analysis is shown in *Table 13.4* where it will be seen that the major cost was on underground repair and that the cycle material cost was 47% of the total cost.

13.5.2   Indirect Cost of Diesel Maintenance

This was associated with the level of equipment unavailability and no attempt had been made in the costing system to quantify it. However, availability figures were recorded for each unit for each operating period and were being used for subsequent planning and budgeting decisions. The availability figures were examined for all diesel units over a two year period and the lowest by far were those of the 5 yd diesel loaders in the copper mining areas. The average true availability (e.g. minus meal breaks) of the 5 yd units was of the order of 60%.

It was obvious even at this early stage of the investigation that this was an unusual situation in that the direct cost was so high and the availability low. In order to establish the reasons for this state of affairs it was felt necessary to:

1. Carry out a more detailed cost, availability and failure analysis of the diesel loaders (thus identifying the main factors causing the high incidence of breakdown).

2. Carry out a more detailed investigation of the organisation and control of the maintenance diesel cycle resources (thus identifying the main factors, if any, causing the inefficient use of such resources).

13.6   Cost, Availability and Failure Analysis of Diesel Loaders

The type of diesel loader first used on production was the 5 yd bucket capacity, Type A, and 13 such units still operated in the copper mining area. The Type As were gradually replaced by Type Bs and Type Cs (a local design of the same bucket size) but it will be seen from *Table 13.3* that the direct costs for these more recent units had not altered to any great extent. It was therefore decided to concentrate on the cost and availability investigation of the Type A diesel units.

A typical cost and availability trend for a Type A unit is shown in *Figure 13.2*. It can be seen that the period maintenance cost rose gradually over the first year of a unit's life and then fluctuated around £0.27/100 t mined before going out of control; in many cases the period cost was found to be out of control for several periods before an overhaul (or extensive repair) was carried out. Overhauling the unit did not bring the maintenance cost/100 t mined back to the as-new level.

Also shown on the graph are the main contributions to the period maintenance costs, for example the failure of major assemblies. It was

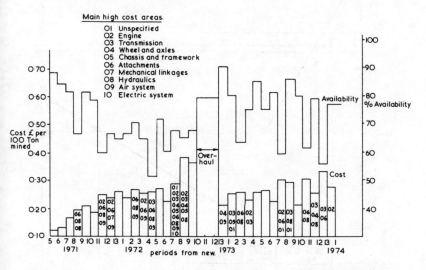

Figure 13.2. Typical maintenance cost and availability trend for a copper mining diesel loader

noticed from the graphs that reconditioned assemblies and sub-assemblies were failing frequently and often shortly after installation. It was therefore decided to investigate the reasons for these failures in more detail and to this end a failure analysis of the Type As was conducted.

The analysis was restricted by the shortage of accurate data but those components that were analysed showed either a 'random' or 'infant mortality' failure mode. An example of such a random failure mode is shown in *Figure 13.3* for the Type As universal couplings and in this case a serious failure was being caused by loose bolts; a simple design modification corrected the fault.

The overall analysis and preliminary on-site investigation indicated that the high incidence of breakdowns appeared to be caused by a combination of the following factors: maloperation of units, the cumulative effect of poor quality repair, the low quality of reconditioned parts.

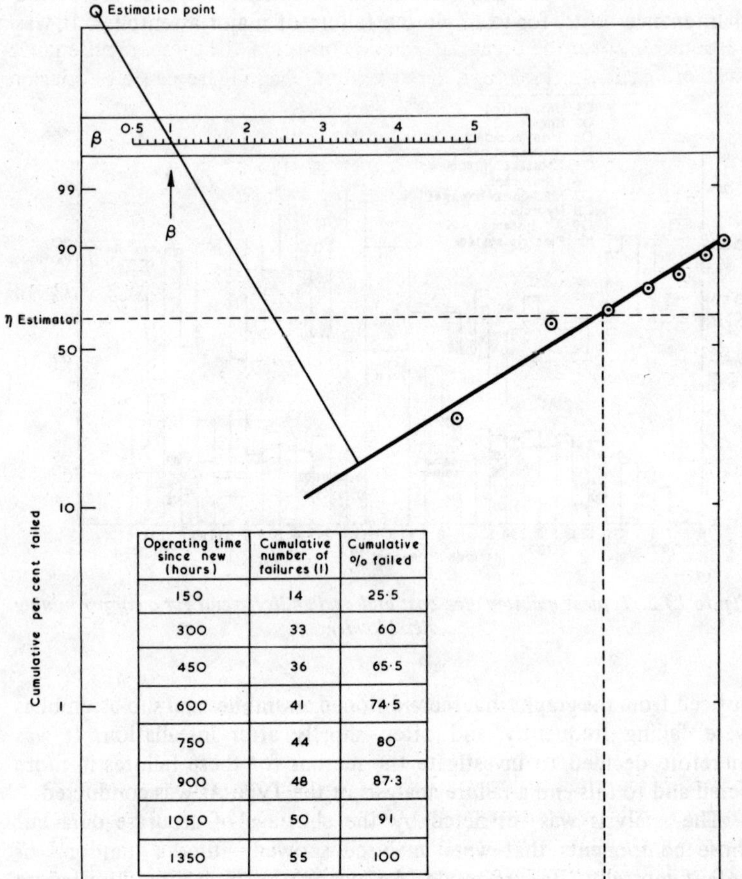

Operating time since new (hours)	Cumulative number of failures (I)	Cumulative % failed
150	14	25·5
300	33	60
450	36	65·5
600	41	74·5
750	44	80
900	48	87·3
1050	50	91
1350	55	100

Figure 13.3. Weibull plot of times-to-failure; universal couplings

13.7 Organisation and Control of Diesel Maintenance System

The surface and the underground diesel maintenance operation were the responsibility of two different engineering departments (*Figure 13.4*) each responsible to the Engineering Director. Thus there was no central responsibility for the organisation and control of the diesel maintenance system, each section to a large extent working separately. Liaison

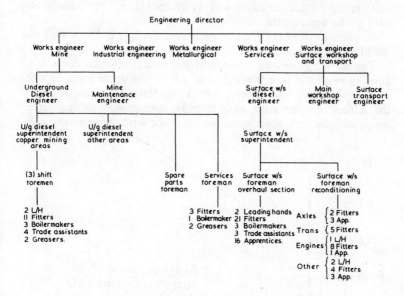

Figure 13.4. Engineering administrative structure

between the engineering sections for decisions concerning overhaul or major repair was through the Diesel Engineers and liaison for the transportation and availability of reconditioned assemblies was through the underground planning foreman and the workshop superintendent.

A procedure for the identification, analysis and rectification of equipment failure problems did not exist. A procedure for the recording and feed-back of failures was in operation both from the underground areas and from the surface workshop. This in theory could have been used in conjunction with the costing system for the identification and analysis of failures. However, this would have been difficult without modification to the system, because inaccurate data were being filed inefficiently, and in a number of places, without cross reference.

13.7.1   Planning of Underground Maintenance

The underground diesel engineer (U.D.E.) had the responsibility for the repair and servicing of all underground diesel equipment; he also had the responsibility of deciding when equipment should be overhauled, and, through his superintendents, when equipment should be transported to the surface workshop for major repair. The working priority for units in the surface workshop was set by the U.D.E. after consultation with production control.

In order to carry out his responsibilities the U.D.E. had a staff and workforce as shown in *Figure 13.4*.

The servicing foreman, in conjunction with production control, set and organised the service programme for all the diesel units. Services on the production and development scoop trams were carried out by the respective repair gangs while all other services were carried out by the small tradeforce directly responsible to the servicing foreman.

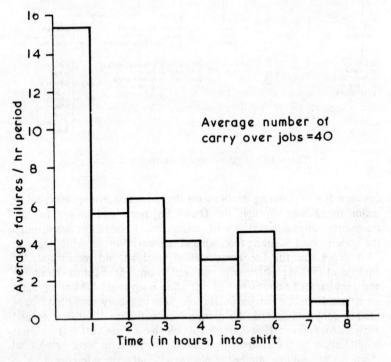

Figure 13.5. Distribution of breakdowns over a shift

The Shift Foremen were responsible for the short term work planning and were aided by the leading hands. At the shift change the oncoming shift foremen were supplied with a known shift work load. The leading hand organised, in conjunction with the spare parts foreman, the parts necessary for the shift work load. On arrival underground the shift foreman decided on the work priority and in this respect there was little communication with production and no written directive. The unofficial priority order was

1. Immobile units.
2. Servicing.
3. Diesel loaders.
4. Small defects.

Each subsequent job was inspected and added to the work load.

Examination of the incidence of breakdowns over a shift revealed a substantial peak of work arriving in the area workshops in the first hour (see *Figure 13.5*). This was caused by the preceding shift operators not reporting faulty units and resulted in planning difficulties and diesel unit queueing early in the shift.

13.7.2   Planning of Surface Maintenance

The surface diesel engineer (S.D.E.) was responsible for the overhaul and major repair of the diesel units and for the reconditioning of the units' assemblies and sub-assemblies. The surface diesel workshop was run as two separate sections, the overhaul/major repair section and the reconditioning section, each section having a foreman responsible for the organisation of labour and for work planning.

13.7.3   Overhaul and Major Repair Section

The work load of this section was evenly divided between overhaul and breakdown work with the overhaul work being concerned mainly with loaders. The work input to this section for a six-month period was analysed and it was obvious that there was some scope for improvement to the forward planning of the repair and overhaul work. The unnecessary randomness of the work input caused large fluctuations in the work load with consequent long unit delays through queueing, for repair.

It was also established that the workforce was inexperienced (only 4 of a section of 29 men had been in the workshop for longer than 6 months) and this in conjunction with a lack of modern production methods and planning, meant that trade-force performance was low and work quality was poor.

13.7.4   Reconditioning Section

It will be seen from *Figure 13.1* that the reconditioning workshop was part of the reconditioning cycle. Its function was to maintain a supply of reconditioned parts to the underground repair shops (95%) and the overhaul section (5%). The total number of reconditioned parts in operation or in the cycle had grown to a level which permitted the workshop to recondition on a batch basis.

The utilisation of the tradeforce in this shop was found to be much higher than that in the overhaul section due to the permanency of the tradeforce and as a result of better short term planning. However, lack of spare parts caused considerable difficulty in maintaining the supply of certain assemblies and also disrupted the work planning. Few jobs had been properly methodised and standard times were not used.

There was no central control of this cycle (the responsibility being divided between underground, surface and warehouse) and no clear record of the total number of reconditioned items in the system, or where these items were located at any particular time; in addition there was no feedback on the scrapping of the smaller items. Thus there were too many of some items in the cycle, with consequent high holding costs and too few of others, with consequent stockout and disruption of the workshop planning.

The other main problem with this cycle was inadequate sub-storage and the extremely ineffective transportation system both to and from the underground workshops. This transportation system was common to the spare parts system.

13.7.5   Spares Organisation and Control

In general the maintenance management considered the supply of spares as ineffective and suggested that this problem caused disruption and serious delays to maintenance work in all parts of the cycle. It was difficult to determine the extent of the problem because of lack of information but it was significant that 'nil stock' often occurred and 12% by value of underground parts were purchased by direct order. The inventory level was about 13% of the capital value of the equipment and this was in addition to the spare assemblies which made up the reconditioning cycle.

Investigation of the initial ordering procedure showed that orders originated at foreman level and were processed through all levels up to Engineering Director before being passed on to the Purchasing Department. In spite of this the *effective* decision was taken at foreman level and often without sufficient information. The re-ordering procedure was complex and did not allow for changing equipment usage rates; communication between maintenance and purchasing was poor.

13.8 Discussion and Recommendations

An outline of the underground diesel maintenance system (or subsystem) is shown in *Figure 13.1*. The system objective was stated by engineering management to be the minimisation of the direct costs of diesel maintenance (through a reduction of 10% per annum) while attempting to achieve a particular level of availability, 70% in the case of the copper mining diesel loaders. From the cost analysis it will be seen that the direct cost of diesel maintenance in order to achieve these levels of availability was about 60% of the capital value of the equipment. Although by normal standards these costs appeared high it was necessary to consider them in relation to the working environment and also to the rate (and value) of production achieved through the use of such equipment. In this respect it must be pointed out that a type of diesel loader operating in the extremely arduous conditions of the copper mining areas could transport £1800 worth of ore each hour. Therefore it was felt that the level of production obtained through the use of diesel equipment (compared to alternative methods) outweighed the disadvantage of even the existing level of maintenance costs.

Having placed the situation in perspective it was still obvious that the direct costs of maintenance were too high and the availability of equipment too low and it was also clear that the solution of the problem was more complex than a mere balancing of the direct costs and availability to achieve a minimum total cost.

The costing analysis showed that the main area of maintenance expenditure was associated with breakdown maintenance in the copper mining area; over half of this expenditure was on material costs. The first and most serious criticism was that no procedure existed for the investigation of breakdowns and high maintenance cost areas. An outline of such a plant condition control system is shown in *Figure 1.4* and it was recommended to management that such a procedure should be adopted if maintenance costs were to be reduced and availability improved. Much of the feedback necessary to operate such a procedure did already exist but was in need of modification; this was especially so in the case of breakdown data feedback. It was also necessary to emphasise that breakdowns were caused by any one of a large number of reasons, e.g. maloperation, poor design, poor maintenance, and it was therefore essential that the investigation team should be inter-departmental and have access to specialist services, e.g. oil spectrometry, where necessary. *The function of such a team would be to establish the cause of breakdown and to determine the best (if any) course of corrective action.*

13.8.1   Maloperation

Many of the diesel loader breakdowns were caused (or aggravated) by maloperation. For example a serious engine breakdown resulting in complete engine replacement, was caused through lack of oil in the air filter. The standard operator-prestart oil-check had not been carried out, resulting in a 'dusted' engine. This was not an isolated incident but little was being done to prevent recurrence. Apart from such incidents it is fair to say that the diesel loaders were being operated under conditions, and in a way, for which they were not designed. Much of the incurred maintenance cost could be justified on economic grounds and the problem was to determine what alterations to the existing operating policies could be made which would achieve a net production/maintenance gain. In this respect the following recommendations were put to management for consideration.

1. Increasing the number of production diesel loaders in the maintenance/production cycle to allow the operation (under most circumstances) of a pool of stand-by units. This would allow pressure to be taken off repair gangs and might result in better quality repairs. In addition production operators would be more inclined to return units for maintenance when a fault had been detected.
2. Increasing the responsibility of operators towards the units by tying units to particular operators on each shift.
3. Improvement in the training of unit operators to enable better pre-start checking and also the earlier recognition of major faults.

13.8.2   Design out

Although such modifications to the production policy as suggested above might well reduce maintenance costs, because of the severity of the operating conditions, the level of unit failures would still be high. It was noticed that many of the existing breakdowns recurred frequently and could be designed out. Although the breakdown report and history record was a suitable vehicle for the identification of recurring failures there was no effective procedure for the determination and justification of design modifications. *It was recommended that such a procedure should be adopted and extended to feed information back to the equipment manufacturer.*

13.8.3 Maintenance

If maintenance costs were to be reduced it was also necessary to consider alterations to the existing diesel maintenance organisation and policies. *In this respect the divided responsibility for diesel maintenance was causing serious organisation and communication problems.* This difficulty together with the lack of central planning of the diesel maintenance cycle was causing serious delays in the repair and overhaul of units. It was recommended that the diesel maintenance system be considered as a separate sub-system of the mining operation and should be the responsibility of one senior engineer. A central planning organisation was also needed for the overall planning, scheduling and control of the diesel maintenance cycle.

Little constructive criticism could be directed towards the location of the workshops, the servicing procedure or the short term planning of the underground repair and service crews (although the priority rules were in need of revision), but it was stressed that serious thought needed to be given to the following.

13.8.4 Repair Policy

The repair policy was based on the assumption that the quickest way of repair was the cheapest, the effective decisions were being taken at shift foreman level. This policy in conjunction with the poor diagnostic skill of the tradeforce (and some of the supervision) and the pressure from production for speed of repair was one of the main causes of high maintenance costs. It was recommended that this area should be investigated to provide a set of clear guidelines for decision making; this would involve repair *vs.* replace decisions for all assemblies and sub-assemblies of the diesel units. In addition a training programme would improve the diagnostic skill of the workforce and supervision. It was pointed out that the effective operation of such a policy with more emphasis on the quality of repair would considerably reduce the need for overhauling units.

13.8.5 Overhaul Policy

From the costing graph (*Figure 13.2*) it was obvious that the overhauling procedure was not effective. In order to improve this situation the following recommendations were put to the management:

1. Identification of units (in particular diesel loaders) in need of overhaul should be through regular *on-site inspection* by a senior diesel engineer. The identification of such units could be aided by the costing and breakdown reports. The condition of the unit should not be allowed to get out of control.

2. The existing procedures for reconditioning assemblies and for overhauling units should be investigated to improve the quality of work. This might bring the overhauled units near to an 'as new' condition and it would be possible to consider overhauling units at the top of their initial cost curve.

13.8.6 Surface Workshop

Traditional production management techniques should be applied to the operation of the surface workshop. In addition the policy of using all the inexperienced tradesmen in the overhaul gangs should be modified in order to establish a nucleus of permanent experienced tradesmen within the overhaul section.

13.8.7 Organisation and Control of Reconditioned and Spare Parts

The spare parts initial ordering procedure should be rationalised and the responsibilities of the personnel involved defined. Decisions involving major assemblies and sub-assemblies should be taken at a senior management level with the necessary information for decision making provided. The re-ordering procedure could be improved by better communication on usage rates between maintenance and purchasing departments. The purchasing department should also investigate further the procedure for reviewing spare parts stock levels. The sub-storage and transportation of spares and reconditioned parts to and from the underground workshop was totally inadequate and had to be improved if the direct and indirect costs of maintenance were to be reduced.

13.9 Conclusions

The observed annual level of direct diesel maintenance cost was about 60% of the capital value of the diesel equipment. Notwithstanding this high level of direct cost the availability of certain critical production diesel units was as low as 60%. Although the profitability of the diesel mechanised method of ore production (when compared to alternative mining methods) can justify high maintenance costs it has been shown that much of the observed costs and unavailability was caused through maloperation, recurring faults, and poor operating policies and was therefore unnecessarily high. *The reduction of the maintenance costs to an optimum level is a dynamic interdepartmental problem the solution of which requires the type of approach outlined in Figure 2.4. This approach also emphasises the need for an effective procedure for incorporating design modifications into future equipment.*

REFERENCES

1. Kelly, A., 'A Case Study of Availability', *Case Study 4*, National Terotechnology Centre (1975)

Index